General Chemistry
Quantitative and Qualitative Laboratory Experiments 3E

Albert E. Russell
Fitzgerald B. Bramwell
Gregory Pritchett
Melissa S. Reeves
Marilyn P. Tourné
Daniel A. Abugri

This edition has been printed directly from camera-ready copy.

Copyright © 2019 by Empire Science Resources, LLC

ISBN 978-1-942465-02-7

All rights reserved. No part of this publication may be reproduced, stored in a retrieval system, or transmitted, in any form or by any means, electronic, mechanical, photocopying, recording, or otherwise, without the prior written permission of the copyright owner.

Printing 10 9 8 7 6 5 4 3 2 1

General Chemistry Quantitative and Qualitative Laboratory Experiments

General Chemistry

Quantitative and Qualitative Laboratory Experiments 3E

**Russell • Bramwell • Pritchett
Reeves • Tourné • Abugri**

Periodic Table of the Elements

1 1A	2 2A	3 3B	4 4B	5 5B	6 6B	7 7B	8 8B	9 8B	10 8B	11 1B	12 2B	13 3A	14 4A	15 5A	16 6A	17 7A	18 8A
1 **H** 1.0079																	2 **He** 4.00260
3 **Li** 6.941	4 **Be** 9.01218											5 **B** 10.811	6 **C** 12.011	7 **N** 14.0067	8 **O** 15.9994	9 **F** 18.9984	10 **Ne** 20.1797
11 **Na** 22.9898	12 **Mg** 24.305											13 **Al** 26.9815	14 **Si** 28.0855	15 **P** 30.9738	16 **S** 32.065	17 **Cl** 35.453	18 **Ar** 39.948
19 **K** 39.0983	20 **Ca** 40.078	21 **Sc** 44.9559	22 **Ti** 47.867	23 **V** 50.9415	24 **Cr** 51.996	25 **Mn** 54.9380	26 **Fe** 55.845	27 **Co** 58.9332	28 **Ni** 58.6934	29 **Cu** 63.546	30 **Zn** 65.409	31 **Ga** 69.723	32 **Ge** 72.64	33 **As** 74.9216	34 **Se** 78.96	35 **Br** 79.904	36 **Kr** 83.798
37 **Rb** 85.4678	38 **Sr** 87.62	39 **Y** 88.9059	40 **Zr** 91.224	41 **Nb** 92.9064	42 **Mo** 95.94	43 **Tc** (98)	44 **Ru** 101.07	45 **Rh** 102.906	46 **Pd** 106.42	47 **Ag** 107.868	48 **Cd** 112.411	49 **In** 114.818	50 **Sn** 118.71	51 **Sb** 121.76	52 **Te** 127.60	53 **I** 126.904	54 **Xe** 131.29
55 **Cs** 132.905	56 **Ba** 137.33	*	72 **Hf** 178.49	73 **Ta** 180.948	74 **W** 183.84	75 **Re** 186.207	76 **Os** 190.23	77 **Ir** 199.22	78 **Pt** 195.08	79 **Au** 196.967	80 **Hg** 200.59	81 **Tl** 204.383	82 **Pb** 207.2	83 **Bi** 208.980	84 **Po** (209)	85 **At** (210)	86 **Rn** (222)
87 **Fr** (223)	88 **Ra** (226.025)	**	104 **Rf** (261.11)	105 **Db** (262.11)	106 **Sg** (263.12)	107 **Bh** (262)	108 **Hs** (265)	109 **Mt** (266)	110 **Ds** (269)	111 **Rg** (273)	112 **Cn** (277)	113 **Nh** (286)	114 **Fl** (289)	115 **Mc** (289)	116 **Lv** (293)	117 **Ts** (294)	118 **Og** (294)

Lanthanoid Series*

57	58	59	60	61	62	63	64	65	66	67	68	69	70	71
La 138.905	**Ce** 140.116	**Pr** 140.908	**Nd** 144.242	**Pm** (145)	**Sm** 150.36	**Eu** 151.964	**Gd** 157.25	**Tb** 158.925	**Dy** 162.500	**Ho** 164.930	**Er** 167.259	**Tm** 168.934	**Yb** 173.04	**Lu** 174.967

Actinoid Series**

89	90	91	92	93	94	95	96	97	98	99	100	101	102	103
Ac (227.028)	**Th** 232.038	**Pa** 231.036	**U** 238.029	**Np** (237.048)	**Pu** (244)	**Am** (243)	**Cm** (247)	**Bk** (247)	**Cf** (251)	**Es** (252)	**Fm** (257)	**Md** (258)	**No** (259)	**Lr** (260)

Standard atomic weights (*Pure Appl. Chem.* 78(11), 2051-2066, 2006 and Aug 2007 release.) Elements with atomic weight in ( ) have no stable nuclides. Well-known isotopes are listed with the appropriate relative atomic weight. Th, Pa, and U have characteristic terrestrial isotopic compositions, and for these an atomic weight is listed. ††IUPAC group labels are in bold type. The smaller group labels below them in regular type are common U.S. usage.

Table of Contents

Preface	5
Disclaimer	8
To the Student*	9
Laboratory Safety*	10
Laboratory Equipment and Techniques*	16
Mathematical Treatment of Data*	29

Experiments

1. Graphing Data with Excel®	36
2. Identification of Compounds Through Physical and Chemical Changes	45
3. Chromatography in Identification of Color Ions and Compounds	55
4. Gravimetric Analysis Water of Hydration and Crystallization*	65
5. Acids, Bases, Salts, and pH	73
6. Stoichiometry of Reaction	87
7. Limiting Reactant	93
8. Standardization of a Sodium Hydroxide Solution	103
9. Volumetric Analysis of Vinegar	111
10. Determination of the Molar Gas Constant R*	117
11. Recognizing the "Fingerprints" of Atoms and Molecules	125
12. Absorption Spectrophotometry and Beer's Law*	137
13. Solution Concentration	145
14. Molecular Geometry	151
15. Synthesis of Alum	165
16. Kinetics of Bleaching a Food Coloring	173
17. pH Measurement and Buffer Solutions*	185
18. Natural Acid-Base Indicators and pH	199
19. Solubility Product*	207
20. Standard State Determination for Urea Solvation	217
21. Galvanic Cells*	225
Introduction to Qualitative Analysis*	237
22. Qualitative Analysis of Cation Group I*	247
23. Qualitative Analysis of Cation Groups II-V*	253
24. Qualitative Analysis of a General Cation Unknown*	267
25. Qualitative Analysis of Common Anions*	273

Appendices

A. Names and Formulas of Common Ions*	281
B. Tables of Data*	282
1. General Solubility Rules for Common Salts and Hydroxides	283
2. Saturated Vapor Pressure of Water from 0° to 100°C	284
3. Relative Density of Water	284
4. Prefixes for Fractions and Multiples of SI Units	285
5. Equilibrium Constants for Acidic and Basic Dissociation at 25°C	285

6.	Solubility Products at 25°C	286
7.	Concentration of Side-Shelf Acid and Base Solutions	287
8.	Standard Reduction Potentials at 25°C	288
9.	Common Buffer Solutions	289
10.	Color Changes, pH Intervals, and H_3O^+ Concentration Intervals of Some Important Indicators	289
C.	Understanding the Spectrophotomer*	290
D.	Quick View of the pH Meter – Voltmeter*	292

* Adopted with permission from *Basic Laboratory Principles in General Chemistry with Quantitative Techniques*, Bramwell, F.B., Dillard, C.R., and Wieder, G.M., (1990) Kendall/Hunt ISBN 0-8403-5654-4.

Preface

This third edition continues and expands upon the laboratory exercises and pedagogic philosophy of *General Chemistry Quantitative and Qualitative Laboratory Experiments*. New features include an updated and revised Laboratory Equipment and Techniques section, selective report questions, prelaboratory exercises, and Further Reading references. Prelaboratory exercises are now an integral part of each laboratory experiment. Thus, this text, like its predecessors, provides qualitative and quantitative laboratory exercises to serve the needs of a one-year general chemistry program. It was written in the belief that laboratory studies are an essential part of undergraduate education. Each experiment has a well-defined objective that underscores a basic chemical tenet while providing a reliable, reproducible and satisfying result. Specifically, students learn how to perform essential laboratory techniques such as weighing, titration, glass-working, and informed calculations based on experimental data. Moreover, professional conduct including approaches to safety rules, chemical disposal and storage, organization, and neatness in laboratory operations are integral to each experiment.

Through the assembly of scientific apparatus leading to the observation of chemical reactions, this laboratory course stimulates an interest in chemical phenomena. The use of "unknowns" and the use of specific laboratory techniques applied to solve practical problems demonstrate the investigative nature of chemistry. Through these laboratory exercises, students learn that even the most precise scientific measurements are subject to uncertainty. Thereby students learn to distinguish between experimental errors, uncertainties, and "blunders." Thus, the importance of error analysis is introduced at an early stage of their scientific training.

The quantitative, qualitative, and synthetic general chemistry laboratory exercises within this manual may be used in an independent laboratory course, separate from lecture, or in conjunction with a variety of textbooks. This manual is designed for an instructor to schedule experiments that meet the demands of many varied and different student groups.

The laboratory experiments within this manual include a wide range of interesting studies in the general categories of basic principles, techniques of separation and identification; moles, and stoichiometry; chemical thermodynamics; electron transfer; acid-base equilibria; kinetics and physical properties of matter; and synthesis and characterization of inorganic compounds and complex ions.

The manual falls into five parts:
1. Introductory material on experimental procedures, laboratory safety, and mathematical treatment of data;
2. Laboratory experiments;
3. Pre-laboratory preparatory material;
4. Appendices;
5. Laboratory equipment and chemical database (instructor's edition only).

Parts of the manual are designed to take advantage of the vastly increased computing power offered by smart phones, computer tablets, and personal computers. For example, we have introduced the treatment of uncertainty and error analysis as an optional exercise in Experiments 10 and 21.

Instructors may choose any suitable sequence of laboratory exercise to fulfill general chemistry course requirements. For example, an instructor may find that the sequence 1, 2, 5, 7, 8, 6, 12, 19 best fits a particular course. By using Experiments 22-25, it is possible to include qualitative analysis or identification of ions without using a formal qualitative analysis scheme. Obviously, many other sequences are possible.

Experiments were selected according to the following seven criteria:
1. Quantitative experiments must give results which are reproducible and for which experimental error is less than 10%.
2. Experiments should be completed in a single three-hour laboratory period. If this cannot be done, the experiment must include operations that logically can be suspended and resumed at a later time.
3. The laboratory overhead (cost of experiments) must allow the engagement of schools with large

enrollments or with limited budgets.
4. Experiments must include relevant safety data in a professional manner. Appropriate web sites provide access to materials safety data sheets, as well as to the disposal or storage of chemical waste.
5. Experiments must promote a sense of inquiry within general chemistry students. Pre-laboratory reports, introductory material and provocative report sheet questions are designed to encourage students to think independently. Standardized report sheets are designed to help organize student thought, tabulate data and arrive at results. Students must think about how the mathematical manipulation of laboratory data lends itself to theoretical interpretation.
6. Experiments must include scientific literature references that allow students to "dig deeper" into relevant material.
7. Experiments must make use of appendices not only as an integral part of procedure and discussion, but also as a valuable reference source for commonly used physical and chemical constants.

Two unusual features of the instructor's edition of this laboratory manual include a chemical resource database and enhanced production adaptability. To assist laboratory instructors and laboratory administrators, in the instructor's manual we included a database of all equipment, supplies, and chemicals used in each experiment, along with a useful internet connection to appropriate service providers. Thus, an instructor can review any experiment, determine if the necessary supplies are on-hand, and if not, be able to review prices or even order such materials directly from a supplier.

To insure fresh and provocative materials, the manual is designed to allow rapid replacement of pre-laboratory quizzes, report sheet questions, and resource databases. Thus, it is possible to order manuals with different sets of pre-laboratory report questions, experiment report questions or with updated resources while keeping the basic laboratory experiment principles intact.

We believe that students who have learned basic laboratory operations in a general course are better able to grasp the significance of principles expressed in advanced courses. This laboratory manual uses the direct involvement of students in apparatus assembly, recording of observations, and reduction of data to useful forms. We believe that this approach to the general chemistry laboratory maintains student interest as well as generates true enthusiasm and appreciation for the rewards and the frustrations of chemical experimentation. It is critical, therefore, that introductory chemistry courses continue to include laboratory studies that emphasize basic principles, illustrate qualitative results, and promote quantitative techniques. Since the theoretical foundations of chemistry lean heavily upon the results of experiments, training in laboratory techniques must begin immediately in an undergraduate educational experience.

Text sections and laboratory exercises designated in the Table of Contents with an "*" were adopted with permission from *Basic Laboratory Principles in General Chemistry with Quantitative Techniques,* Bramwell, F.B., Dillard, C.R., and Wieder, G.M., (1990) Kendall/Hunt ISBN 0-8403-5654-4.

The intellectual underpinnings of the third edition are based on the helpful comments and criticisms from colleagues at the University of the Virgin Islands, the State University of New York at Old Westbury, Tuskegee University, Brooklyn College of the City University of New York, and the University of Kentucky.

Our greatest accolades, however, go to the many generations of general chemistry students whose feedback and interest continue to be the major impetus for this work.

Disclaimer

The information contained in this laboratory manual is believed to be accurate and represents the best information currently available to us. However, we make no warranty express or implied, with respect to such information, and we assume no liability resulting from its use. Users must make their own investigations to determine the suitability of the information. In no event shall Empire Science Resources, LLC or the authors of this text be liable for any claims, losses, or damages of any third party or for lost profits or any special, indirect, incidental, consequential or exemplary damages, howsoever arising, even if Empire Science Resources, LLC or the authors of this text have been advised of the possibility of such damages.

The safety information contained in this manual has been compiled from sources believed to be reliable in an effort to provide safety guidelines for use in academic laboratories. It is intended to serve only as starting point for good practices. It does not purport to specify minimal legal standards, or provide sufficiency of information. This section is intended to provide guidelines for safe practices; therefore, it cannot be assumed that all necessary warnings and precautionary measures are contained in this document, or that other or additional information or measures may not be required. Users of this manual should consult pertinent local, state and federal laws and legal counsel prior to initiating any safety program.

As a safety precaution, final permission should be given by your instructor to proceed with an experiment for which you have set up the apparatus, in which you are about to participate, or have devised details of procedure.

To the Student

The chemistry laboratory is a place where you can examine, test, and establish for yourself the principles learned in lecture and from your textbook. You will also learn techniques of safe and proper handling of chemicals, glass-working, and accurate measurement. A thorough knowledge of the basic principles and quantitative techniques described in this manual will prepare you to approach scientific problems analytically and efficiently.

The *Laboratory Equipment and Techniques* section presents special laboratory procedures such as glass working and the handling and transfer of reagents and solutions. Read through these procedures before the first laboratory meeting. Later, when you perform specific operations, refer back to the appropriate material.

To get the most out of the laboratory, it is important to be as much concerned with *how* each experimental result is obtained as with its numerical results. Think of each experiment as an investigation of the unknown. Even when the stated objective is obvious, you will find that you will learn something new; if you do not learn new facts, you will learn a new way of thinking about facts that are already known. Therefore, try to approach each experiment in the following systematic manner.

1. Read through the assigned experiment before coming to the laboratory. Note carefully any special techniques, precautions, or warnings, and try to visualize in advance how you will perform each operation. Make an outline of the experimental steps.
2. Look over the report sheet and questions in advance so that you will know what principles are being demonstrated and what data will be required. When you perform the experiment, record the data in ink in the appropriate places on the report sheet rather than on scraps of paper, which may get lost. If a data entry is erroneous, draw a single line through it and place the corrected value next to it. All reports must be submitted in ink.
3. Pay special attention to any sample calculations in the introductory section of the experiment. To perform chemical calculations, you should already be familiar with 'the use of logarithms, exponential notation, and intermediate algebra. Review these topics in other books if necessary. Scientists use proper units and an appropriate number of significant digits to express the numerical results of experiments. Before the first laboratory period, read the section called *Mathematical Treatment of Data* to review the use of significant digits and the factor-label method.
4. Calculate your results before leaving the laboratory, preferably as soon as you have obtained sufficient data. By so doing you can check a measurement, collect additional data, or, if necessary, repeat an experiment.
5. Increase your understanding of the experiment by reading the references cited under *Further Reading.* These references provide additional valuable information and suggest ways of performing the experiment more efficiently. Working responsibly and safely in a chemistry laboratory is a rewarding experience. However, unless you are aware of potential dangers, serious accidents can occur. Carelessness and frivolity can create hazardous conditions endangering you and your classmates. Read the discussions on laboratory safety and observe the safety rules at all times. *Do not attempt unassigned experiments and do not make unauthorized changes in the assigned experiments.*

Laboratory Safety

Most laboratory accidents are caused by negligence. An accident could be your fault or that of the student next to you. Safety in the laboratory is everyone's responsibility. Safety is best maintained by *protection* - of the body against potentially dangerous laboratory situations such as fumes, fires, corrosive or poisonous chemicals, *prevention* - by acting in a manner that minimizes the possibility of accident or injury, and knowledge of the *emergency procedures* - to follow in case of an accident. If there is danger that an accident might occur which would involve you or a fellow student, immediately call your laboratory instructor for assistance.

The probability that you will be the victim of other students' mistakes can be minimized if you all observe the following rules of safety.

Protection

1. *Wear safety goggles in the laboratory at all times.* Do not wear contact lenses in the laboratory. Corrosive chemicals and gases may become entrapped between the lens and your eye, and eye damage may result. Furthermore, when lenses are in place, they cannot be easily removed should you need to wash your eye.
2. *Keep long hair tied back* so that it cannot fall into a flame or into chemicals. Beards, too, may be a hazard. Sprayed hair is particularly dangerous because it presents a large surface area of flammable plastic which will burn rapidly if ignited.
3. *Wear comfortable, closed, low-heeled shoes.* Sandals should not be worn because of inadequate protection for your skin.
4. *Wear simple clothing.* Do not wear scarves, neckties, or anything that may hang into flames or chemicals or get caught in equipment. Natural fibers, which are not as flammable as synthetics, are preferred. Short pants or miniskirts should not be worn because of inadequate protection for your skin. A lab coat or apron offers some protection for you.
5. *Locate laboratory safety equipment.* Do this during the first laboratory period. Know where to find the safety shower, the fire extinguisher, sand pail, fire blanket, and the eye wash fountain. The safety shower is used when a corrosive chemical is spilled on a person in quantities that cannot be flushed away at the laboratory faucets.

Prevention

1. *Prepare well for laboratory assignments.* Study the experiment and relevant techniques *before* the laboratory period. Be aware of possible hazards, concentrate on what you are doing, and follow directions.
2. *Never work alone in the laboratory.* In the event of an accident there would be no one to help you.
3. *Do not attempt unauthorized experiments!* Experiments which are not a part of the regular class assignment may be performed only with the written permission of the instructor.
4. *Clean up the space you occupy before you start working.* Check especially the drain and sink associated with your laboratory space to be sure that no solid materials (matches, paper, etc.) have been left there. Laboratory work areas must be free of clothing and other personal possessions to minimize fire hazards.
5. *Put away all equipment and clean the top of the desk and hood area at the end of the laboratory period.* Check both the drain on top of the desk and the sink at the end of the desk to be sure no solid materials are left there. These are your responsibilities - measures you must take to protect yourself - whether or not the waste on the desk or in the sink was left by a previous class,
6. *Read the labels on reagent bottles carefully.* Use of the wrong reagent may be disastrous. Do not use more, or less, than the amount specified. Dispense reagents at the side shelf and *never* take the bottles to your desk.

7. *Protect reagents from contamination.* Do not put droppers, spatulas, or objects of any kind into a reagent bottle. Do not lay the reagent side of a stopper down on any surface. Recap bottles after use. Try to avoid taking a large excess of reagent, However, if you do take more than you need, never return the excess to the bottle.
8. *When carrying a large reagent bottle,* always use two hands with one hand supporting the bottom of the bottle.
9. *Be cautious of hot glass. Glass has the same appearance whether it is hot or cold.* To prevent painful burns, place all hot glass aside on wire gauze until it is cool.
10. Only Pyrex glass and porcelain containers may be heated safely. Other types of glass may shatter under heat treatment. (Calibrated equipment such as graduated cylinders, burets, pipets, and volumetric flasks must *never* be heated.)
11. *When heating a container, do not use a larger or hotter flame than necessary or leave the heating set-up unattended.*
12. *Fire-polish glass tubing.* The razor-sharp ends of glass tubing can cause serious cuts. Therefore, tubing must be fire-polished after being cut. (See the material on glass-working in the Laboratory Equipment and Techniques section.) Do *not touch the heated part of the tube* until it has cooled to room temperature!
13. *Use* a *lubricant when inserting glass tubing into rubber stoppers.* Make sure that the ends of the tubing are fire-polished and cool. Wet the tubing with water or glycerol and hold it with a towel to prevent injury in case the tube breaks. To start, hold the tubing with a towel about 1 cm from the end to be inserted and insert it into the hole of the stopper with a gentle twisting motion. *Protect your hands with* a *towel.* Keeping your grasp no more than 1 crn from the stopper, gradually twist the tube through until enough protrudes from the other end to allow you to *pull* the tube through the rest of the way. Pulling is the recommended technique to avoid tube breakage and potentially serious injury. 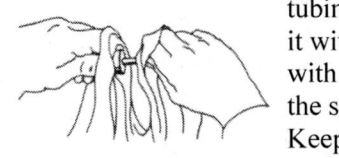
14. *Use* a *lubricant when removing glass tubing from rubber stoppers.* Work a small amount of water or glycerol in between the stopper and the glass. Wrap the glass tubing with a towel *and pull* it from the stopper with a *gentle twisting motion.* 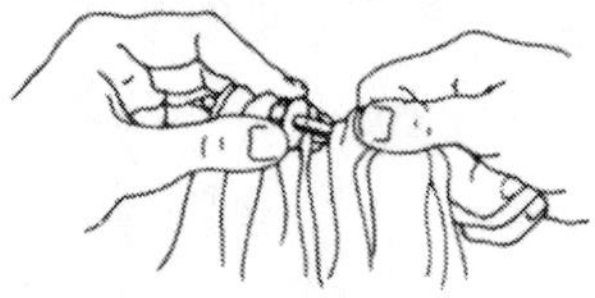
15. *Never look down into* a *test tube or beaker in which* a *reaction is proceeding.* The contents may explode in your face! Also, never point at yourself or others a test tube that is being heated or in which a reaction is taking place.

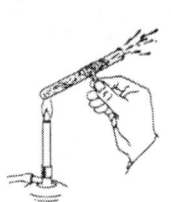

16. *Smell a substance by fanning its vapor gently toward you.*

17. *Never taste anything that is involved in an experiment.* Mouth suction should never be used to fill pipets, to start siphons, or for any other purpose. If you should accidentally get a chemical into your mouth, rinse it out immediately with copious quantities of water, and then inform your instructor. Don't bring food to the laboratory. *Never smoke in the laboratory.*

18. *Never pour water into concentrated acids or bases.* The heat of the reaction may be so great that spattering may occur. Instead, *pour the acid slowly into water with constant stirring.*

19. *Always use the hood when working with irritating or harmful fumes.*

Laboratory Safety

20. *Do not place equipment near the edge of the laboratory bench or too close to your neighbor's equipment.*

21. *Set up your equipment in a professional manner.* Unstable set-ups invite accidents. Do not use broken or defective equipment.

22. *Dispose of wastes and hazardous substances in a manner designated by your instructor.* Improper disposal of chemicals may create a danger to others working in the laboratory. Non-degradable organic compounds and certain heavy metals require specific methods of disposal such as those suggested in the *Safety and Disposal* section of each experiment.

Emergency Procedures

1. *Report to the instructor any injuries or accidents that occur.* Flush slight burns with *ice cold water* for several minutes. Wash cuts thoroughly with soap and water, rinse with clean water, and have the instructor apply a temporary bandage. If you get something in your eyes, immediately flush them out with large amount of water and call for assistance from your laboratory instructor.

2. A small fire on the bench can be smothered by covering it with sand. Act quickly so that the fire cannot spread. Towels should not be used for smothering because they can catch fire themselves.

3. An open fire in the laboratory may be extinguished by discharging a fire extinguisher aimed at the base of the flames and directing the spray from side to side.

4. *Neutralize spilled acids and bases immediately.* Use sodium bicarbonate for acids and dilute acetic acid for bases. After neutralization, remove the residue with water and wipe dry.

5. *Clean up all spills and breakage fragments promptly.* Consult your instructor about the proper procedure if necessary.

Name _____ Lab Section _____ Date_____

Quiz on Safe Procedures for the
General Chemistry Laboratory

1. What must be worn at all times in the laboratory to decrease the risk of eye injury?

2. Answer the following questions on eye protection:
    (a) What device do you use if a chemical gets into your eyes?

    (b) Where is this device located in your laboratory?

    (c) Explain how this device should be used for maximum effectiveness?

3. Why should no one be permitted to work alone in the laboratory?

4. Answer the following questions on personal practices in the laboratory:
    (a) Why must the laboratory work areas be free of clothing and other personal possessions?

    (b) Why is it unsafe to bring food to the laboratory?

5. Answer the following questions on laboratory emergency equipment:
    (a) What should be used immediately for a large chemical spill on clothing?

    (b) Where is this device located in your laboratory?

6. How can a small fire on the laboratory bench be put out safely?

7. Answer the following questions on fire equipment:
    (a) Describe how to extinguish an open fire in the laboratory.

    (b) Where is the fire equipment located in your laboratory?

8. Why must stirring rods and cut-glass ends be fire polished?

Laboratory Safety

9. Describe a safe procedure for smelling a substance.

10. How do you protect yourself and others against poisonous, irritating, or flammable vapors?

11. How do you protect your hands when inserting glass tubing into rubber stoppers or removing glass tubing from rubber stoppers?

12. What precautions are needed with long hair, beards, scarves, or ties in the laboratory?

13. How do you dispose of chemical solids and organic liquids in the laboratory?

14. How may concentrated acid or concentrated base be safely mixed with water?

15. What precautions must be taken when heating a substance in a test tube?

16. Why must reagents be added cautiously?

17. Answer the following questions on laboratory clothing:
    (a) What types of footwear should be worn when working in the laboratory?

    (b) Why are high heels or open toed shoes not acceptable?

18. What types of clothing should you wear in the laboratory?

19. What immediate action should you take for small chemical spills on your skin or clothing or for minor bums?

20. Why is it important to clean the laboratory work areas thoroughly before starting an experiment and after it is completed?

COMMENTS

Laboratory Equipment and Techniques

Equipment and techniques that you will use frequently in your laboratory work are described in this section. Fig. LT-1 shows common laboratory equipment with which you should be familiar, although not all of these objects are discussed in detail here. Read this section carefully before the first laboratory meeting and refer to it as necessary throughout the course.

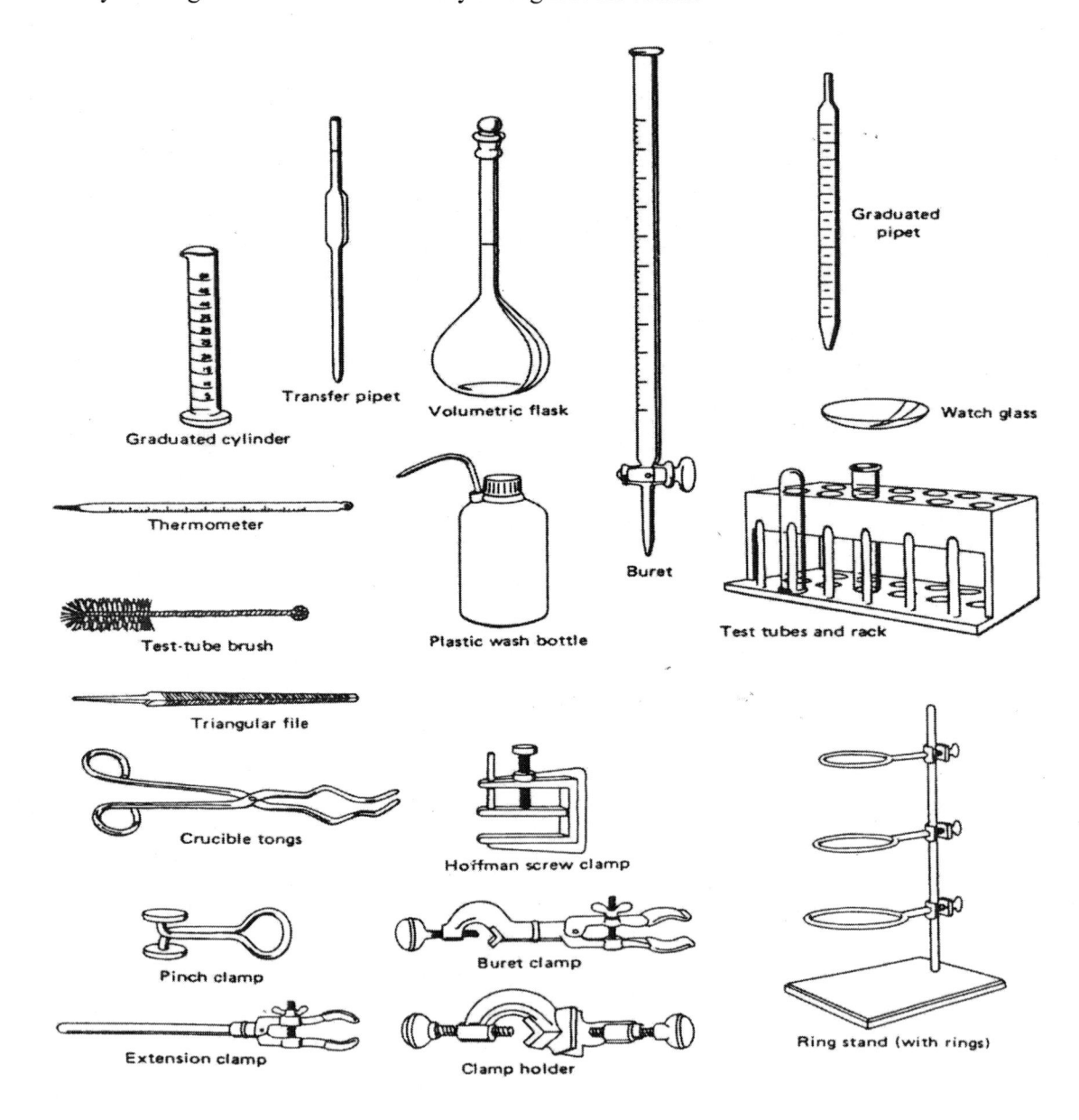

Fig. LT-1 Common Laboratory Equipment

Laboratory Equipment and Techniques

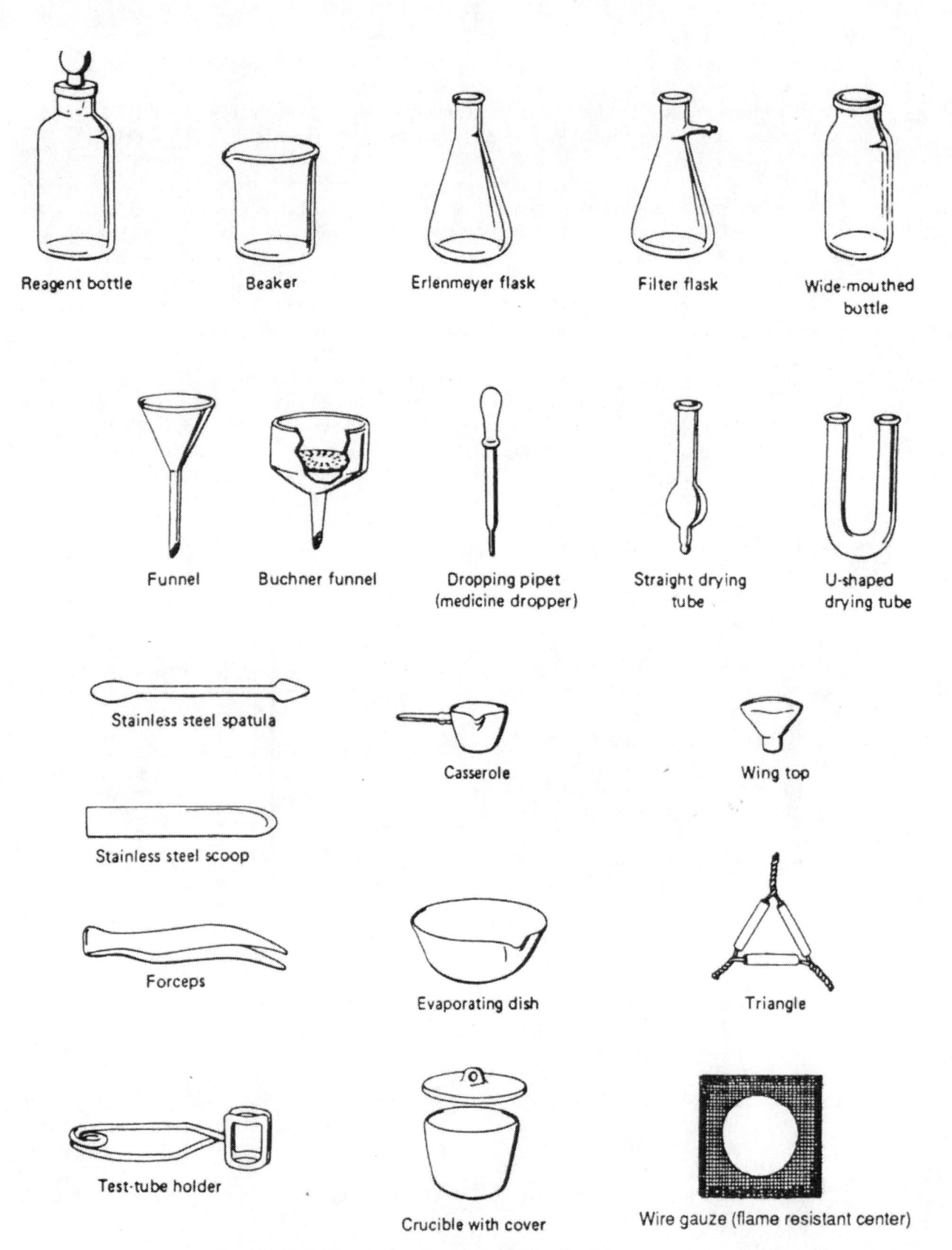

Reagent bottle | Beaker | Erlenmeyer flask | Filter flask | Wide-mouthed bottle

Funnel | Buchner funnel | Dropping pipet (medicine dropper) | Straight drying tube | U-shaped drying tube

Stainless steel spatula | Casserole | Wing top

Stainless steel scoop

Forceps | Evaporating dish | Triangle

Test-tube holder | Crucible with cover | Wire gauze (flame resistant center)

Fig. LT-1 continued. Common Laboratory Equipment

Laboratory Equipment and Techniques

THE BUNSEN BURNER

The basic features of a Bunsen burner are shown in Fig. LT-2. Gas is introduced into the burner through a valve below the air vents by a rubber hose connected to the gas supply. Turning the valve on the burner adjusts the height of the flame. The temperature of the flame can be controlled by adjusting the amount of air supplied to the burner. Try the following experiments with your burner.

1. Attach the burner by means of rubber tubing to a gas outlet and turn the collar at the base of the burner so that the holes are closed and no air can enter. Close the burner gas valve completely and then open it one revolution. Turn on the gas supply knob fully and light the gas by bringing a lit match to the side of the top of the burner. Observe the character of the flame produced. This is called a luminous flame. Using crucible tongs, hold a cool porcelain dish in the flame for about 20 seconds and then look at the surface that has been exposed to the flame. What is the deposit? The luminous flame is not very hot and does not provide efficient heating.

2. Turn the collar at the base of the burner so that air can enter and mix with the gas. Use the valve at the base of the burner to regulate the supply of gas so that the flame is about 4 inches (10 cm) high, and adjust the air openings so that all luminosity is gone and two zones appear in the flame. The approximate temperatures reached by the various portions of the flame by observing a clean piece of wire gauze change color in the flame. A color scale is given in Table LT-1. With crucible tongs, hold a piece of wire gauze horizontally in the flame at different heights. Indicate on Fig. LT-2 how the temperature varies in different parts of the flame. In what areas of the flame are the highest and the lowest temperatures reached?

If the collar at the base of the burner is turned so that the ratio of air to gas is too high, the flame may "strike back" and burn inside the barrel. Then the barrel of the lighted burner gets hot, there is a rasping sound, and the flame may become small, slender and show no cone-shaped regions. If this happens, turn off the gas, let the burner cool, close the air vent, and relight the burner. A burner which "strikes back" easily is probably defective and should be exchanged for a new one.

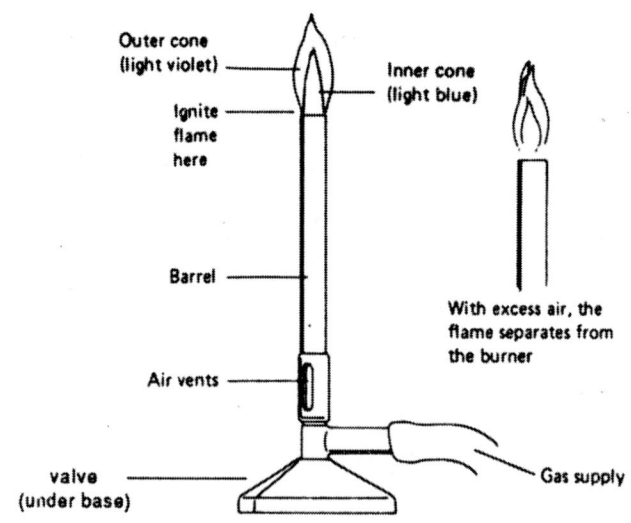

Fig. LT-2. Bunsen burner.

Temperature (°C)	Color of Wire Gauze
500	incipient red
700	dark red
900	bright red
1100	orange
1300	incipient white
1500	white

Table LT-1. Color and Temperature of Wire Gauze.

Laboratory Equipment and Techniques

GLASS WORKING

Caution: glass has the same appearance whether it is hot or cold. To prevent painful burns, place all glass with which you have been working on wire gauze until it is cool. Do not attempt glass-working operations near open windows. The draft from the window may cool both the glass and the flame to temperatures at which the glass is too hard to be workable. Do not put hot glass into wastebaskets - it might cause a fire.

1. *Cutting glass tubing.* With a single stroke of a glass cutter or a triangular file, make a scratch across the glass at the desired place. Put both thumbs close together on the side of the tubing opposite the scratch and, with the scratch pointed away from you, apply pressure with your thumbs and break the glass (Fig. LT-3). Do not twist the tubing. If gentle pressure does not produce the desired result, make the scratch slightly deeper and repeat the operation. Occasionally, tubing does not break evenly. To correct this, hold the tubing more firmly; then, in one motion, pull and snap (but do not twist) the tubing.

2. *Fire-polishing glass tubing, stirring rods.* The most common student accident is a cut from the razor-sharp ends of glass tubing. All tubing must be fire-polished after being cut. Hold the sharp edges of the tubing in the flame and rotate the tubing until the flame is bright yellow. Stop heating as soon as the edges are rounded because overheating the tube will cause it to collapse. Let the tubing cool on wire gauze before you handle it again. Get a piece of solid glass rod 3- to 5-mm in diameter from the storeroom and make two stirring rods, each about 5 inches (120 to 135 mm) long; fire-polish both ends of each rod and allow the rods to cool on wire gauze.

3. *Bending glass tubing.* Place a wing top on the burner to spread the flame; this makes it possible to heat a greater length of tubing. Hold a 25-to 30-cm piece of tubing lengthwise in the upper portion of the flame and rotate the tubing steadily until the glass softens. Then remove it from the flame, bend it to the desired angle, and let it cool on a piece of wire gauze (Fig. LT-4). Save this bend for future use. *Caution: be sure that glass that has been heated has cooled before you handle it.*

4. *Drawing out glass tubing: constructing* a *tip.* Heat a small section of a short length of tubing by rotating it in a flame without a wing tip. When the section is soft, remove the tubing from the flame and draw the ends apart until the diameter is reduced to the desired size. Continue to rotate the tube while drawing it out.

   Prepare a capillary dropping pipet from a piece of 7-mm glass tubing. The capillary portion should be at least 5 cm long and the wide portion about 8 cm long, as shown in Fig. LT-5.

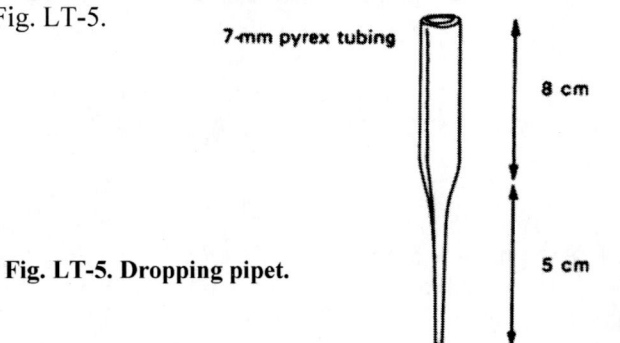

Fig. LT-3. Breaking glass tubing.

Fig. LT-4. Bending glass tubing.

Fig. LT-5. Dropping pipet.

Laboratory Equipment and Techniques

THE CRUCIBLE

The porcelain crucible used in the general chemistry laboratory is a relatively fragile piece of apparatus that is rated to withstand intense heat. Avoid touching cleaned crucibles with your hands. Use a lint-free towel or crucible tongs (Fig. LT-6) to transfer crucibles from the laboratory workbench to a balance or other location. In this manner crucibles can be moved from one place to another without becoming contaminated by oils present on your fingers. Never hold the lip of a crucible with the tongs - it might shatter.

Fig. LT-6. Using tongs to transfer a crucible.

THE DIGITAL BALANCE

Digital balances of the type used in the general chemistry laboratory provide a precision of 0.01 g. The top of the scale is where you place your material for weighing. It is usually flat and made from stainless steel. To protect this surface, it is best to measure material using a beaker, flask or weighing paper. In the front of the scale, you will find a digital readout that will display the mass of whatever is placed on it. Depending on your preference, the readout will display the total mass, the tare mass, the net mass, or a combination of all three.

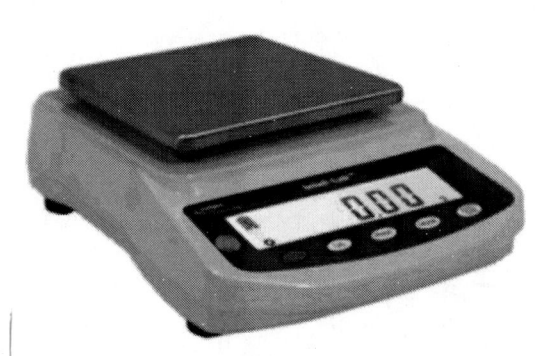

Fig. LT-7. Top loading digital balance

Digital scales use a strain gauge to distribute evenly the mass of whatever is placed on the top of it, and transfer the evenly distributed mass to a load cell. The load cell, which is part of an electrical circuit, bends as weight is applied. The electrical resistance changes when the load cell bends. The change of this electrical resistance is passed through a voltage converter that sends it to a small microchip responsible for the numbers that appear on the scale display. The instructor will give you specific details for the model used in your laboratory.

GLASSWARE

The glassware in a general chemistry laboratory is used mainly for containing reagents or for providing specific volumes of liquids. Examples of reaction vessels are beakers and flasks. They are generally constructed of Pyrex glass, a material which does not easily shatter when heated, even with an open flame.

Graduated cylinders and volumetric flasks are examples of glassware designed to *contain* a specific volume of a liquid. Transfer pipets and burets are examples of volumetric glassware designed to *deliver* a specific volume of liquid to another container (e.g., a beaker or flask). Transfer pipets are often referred to as volumetric pipets. Volumetric glassware is generally calibrated for a specified volume of liquid for use at a specific temperature. Graduations or calibration marks etched on the surface of the glassware denote the volume it contains or delivers. Volumetric glassware must not be heated even if it is made from Pyrex. Heat may distort the glass sufficiently to render the vessel useless as a calibrated volumetric container. Moreover, the uneven distribution of heat may cause the vessel to crack along one of its graduation marks.

Cleaning Glassware

The inner walls of all glassware must be free of grease and dirt. To test for cleanliness, rinse the glass vessel with water, wipe the outside dry, and watch the drainage from the inner walls. Water will drain uniformly from a clean glass surface but will bead up on a dirty surface (Fig. LT-8).

Laboratory Equipment and Techniques

The cleaning agent must be removed by thorough rinsing with tap water followed by rinsing with distilled or deionized water. Do not use a test tube brush to clean volumetric glassware because scratches may alter the calibration.

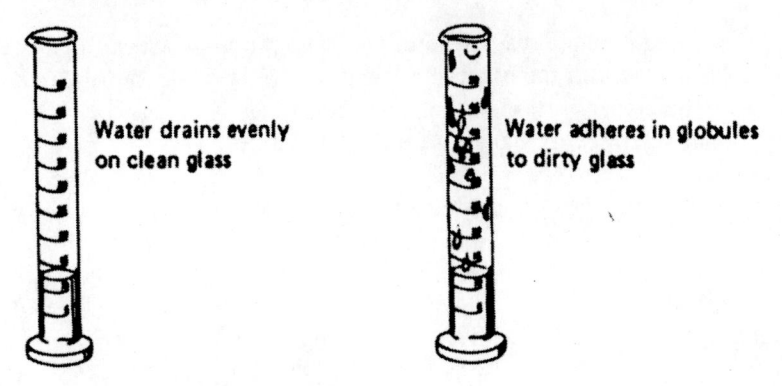

Fig. LT-8. Clean and dirty glassware.

Reading Volumes in Calibrated Glassware

An aqueous solution in a glass container forms a concave surface called a meniscus. Fig. LT- 9 shows how to read the position of the bottom of the meniscus of a clear liquid in calibrated glassware. The apparent change in volume of a solution in calibrated glassware that results from viewing the meniscus at different angles is called *parallax*. You can avoid systematic errors caused by parallax by placing your eye exactly on a level with the meniscus (Fig. LT - 9). When reading the volume levels of intensely colored solutions, such as $KMnO_4$, you may *not* be able to read the bottom of the meniscus. Only in such cases, read the volume of the solution as the *top* of the meniscus. When the meniscus falls between two graduation marks, its position is read as the lower mark plus one estimated digit. For example, the volume of liquid in a 50-mL graduated cylinder can be read to the nearest 0.2 mL if the smallest calibration marks correspond to 1-mL intervals.

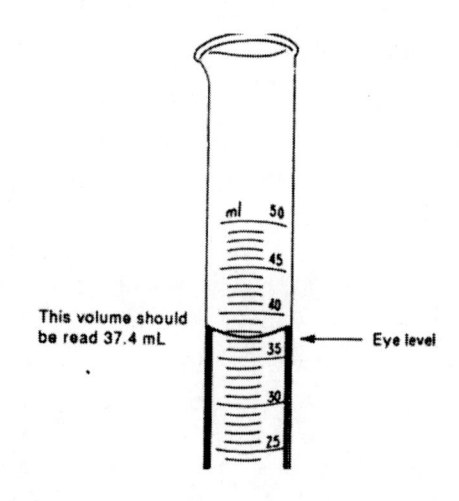

Fig. LT-9. Reading calibrated glassware.

Using Burets

A buret is designed to deliver a precise volume of liquid to a container (beaker or flask). Handle burets with care because they are carefully calibrated and rather expensive.

Liquids are dispensed from the buret through a stopcock which consists of a barrel fitted with either a Teflon or a glass plug containing a narrow channel. The stopcock controls the rate at which liquid is dispensed from the buret. When the channel bore is perpendicular to the body of the buret, the stopcock is completely closed (Fig. LT-10). Teflon plugs need no lubrication. However, those made of glass require a thin, uniform film of stopcock grease to prevent leakage of liquid around the barrel.

Burets should always be clamped in a vertical position to a ring stand or a buret stand. Test burets before use for leakage, uniform drainage, and clean them if necessary. If a buret is dirty, drops of solution adhere to the inner surfaces, and the volume delivered will be in error. After any necessary cleaning, flush the buret body and tip first with tap water and then with distilled water in order to remove the detergent or cleaning solution used. A pipe cleaner is helpful in removing grease or blockages from the plug bore or the buret tip.

Laboratory Equipment and Techniques

Next, rinse the buret walls and tip with two or three 5-mL portions of the solution that is to be delivered. This will avoid dilution by any distilled water remaining on the inner walls. The outside of the buret should be wiped dry so that only liquid from the interior can flow into the receiving vessel.

The buret may now be filled with the reagent solution. Open the stopcock slightly to allow a small amount of solution to run through and fill the buret tip. Any air in the buret tip must be removed and the solution meniscus adjusted to the upper portion of the buret scale (usually slightly below the zero mark). Use a clean stirring rod to remove any drops hanging from the buret tip.

The buret is now ready for use. Merely opening and closing the stopcock will allow a desired volume of solution to enter a collection vessel. To avoid loss of material when delivering liquid, keep the tip of the buret below the lip of the collection vessel (Fig. LT-10). If a flask is used as a collection vessel, gently swirl the contents while the solution is being delivered from the buret (Fig. LT-11). If the collection vessel is a beaker, stir the contents with a clean stirring rod.

When the desired volume of solution has been delivered, any drop hanging on the buret tip should be removed by contact with the inner wall of the collection vessel. The volume delivered from the buret is the difference between the final and initial buret readings. In a titration against a standard solution, the volume delivered is also called the titer volume.

Before reading the meniscus of the solution level, allow 15 to 20 seconds for the solution to drain down the buret's inner wall. If the smallest calibrated divisions correspond to 0.1 mL in the buret, volumes can be estimated to the nearest 0.02 mL. Again, your eye must be at the meniscus level to avoid errors caused by parallax. The meniscus will appear sharper if a card or paper with a darkened area is held behind the buret (Fig. LT-13). After use, the buret, including the tip, should be thoroughly rinsed with water and the outside dried. Instructions on storing the buret will be given by your instructor.

Stopcock barrel — **Closed plug**

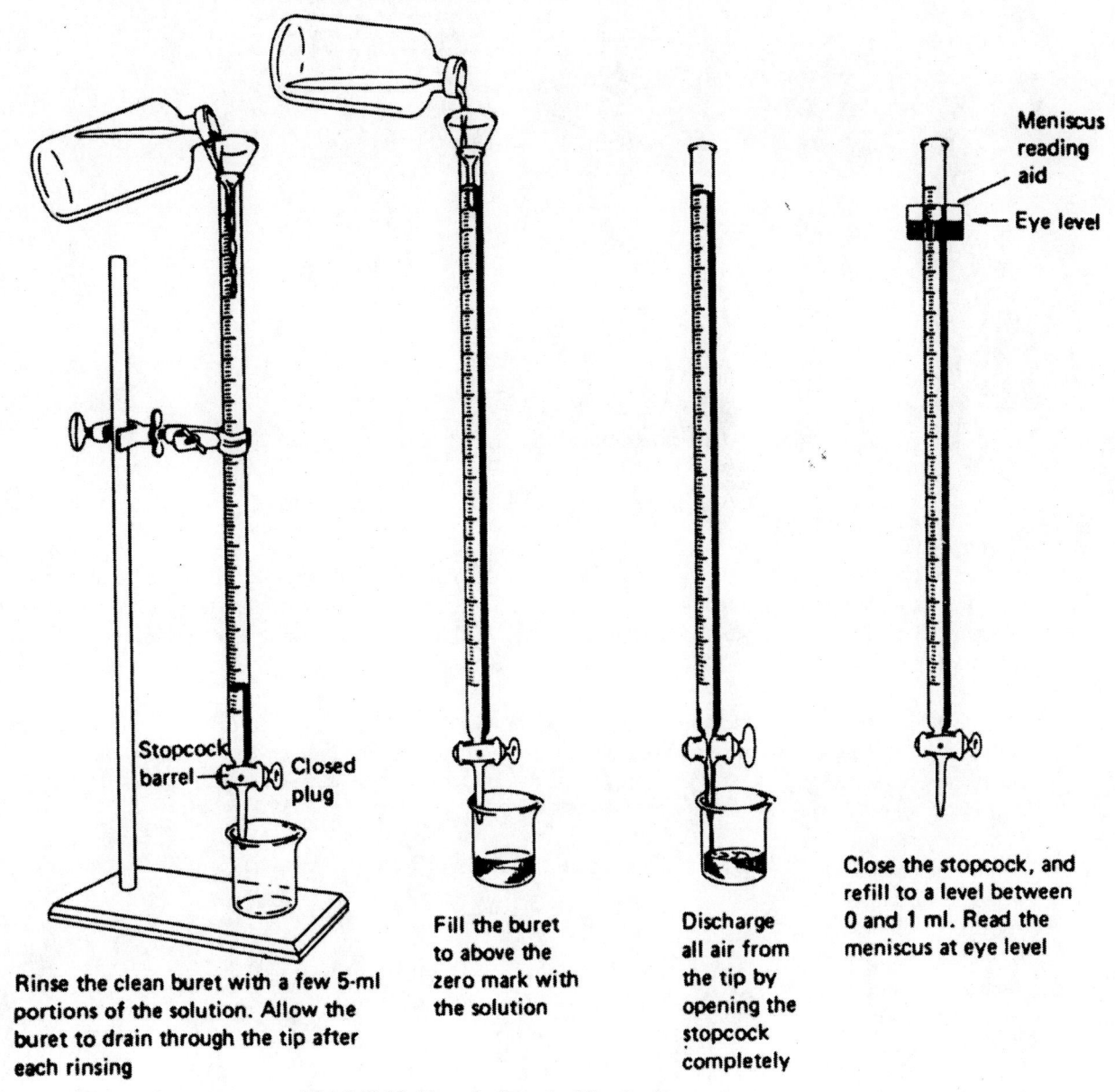

Rinse the clean buret with a few 5-ml portions of the solution. Allow the buret to drain through the tip after each rinsing

Fill the buret to above the zero mark with the solution

Discharge all air from the tip by opening the stopcock completely

Meniscus reading aid — **Eye level**

Close the stopcock, and refill to a level between 0 and 1 ml. Read the meniscus at eye level

Fig. LT-10. Use of a buret with a beaker.

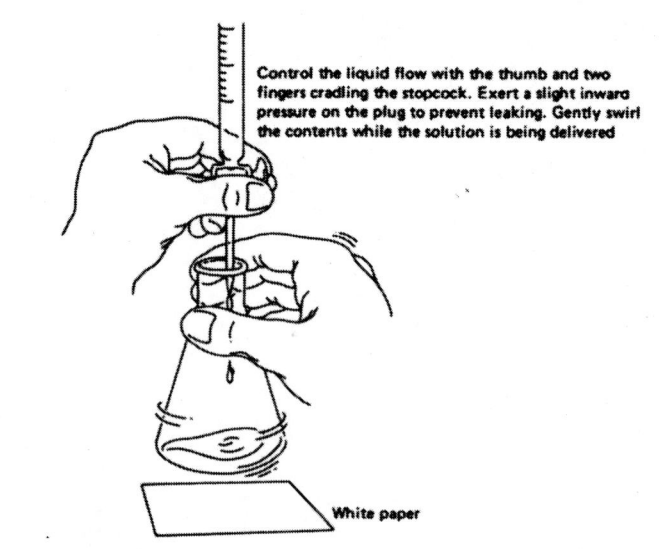

Fig. LT-11. Use of a buret with a flask.

Using Pipets

The transfer (volumetric) pipet is calibrated to deliver a precisely known volume of liquid at a specified temperature from one container to another (Fig. LT-12). The particular temperature and volume are marked on the barrel or stem. An accuracy of 0.2 of the specified volume can be achieved with the transfer pipet.

To fill a pipet, apply suction at the upper stem while the tip is immersed in liquid. Obviously, *you should not use your mouth* to supply the suction since many solutions are corrosive or poisonous. Do *not pipet from reagent bottles.* Pipet only from temporary storage containers such as beakers, storage bottles and flasks. Generally, a rubber pipetting bulb, vacuum, or aspirator is used to provide the necessary suction. See Fig. LT-12 for detailed instructions on how to use a transfer pipet with rubber bulb suction. As with other forms of glassware, you can test a pipet for cleanliness by rinsing it with distilled water. If drops adhere to the inner walls upon drainage, you must clean the pipet before using it.

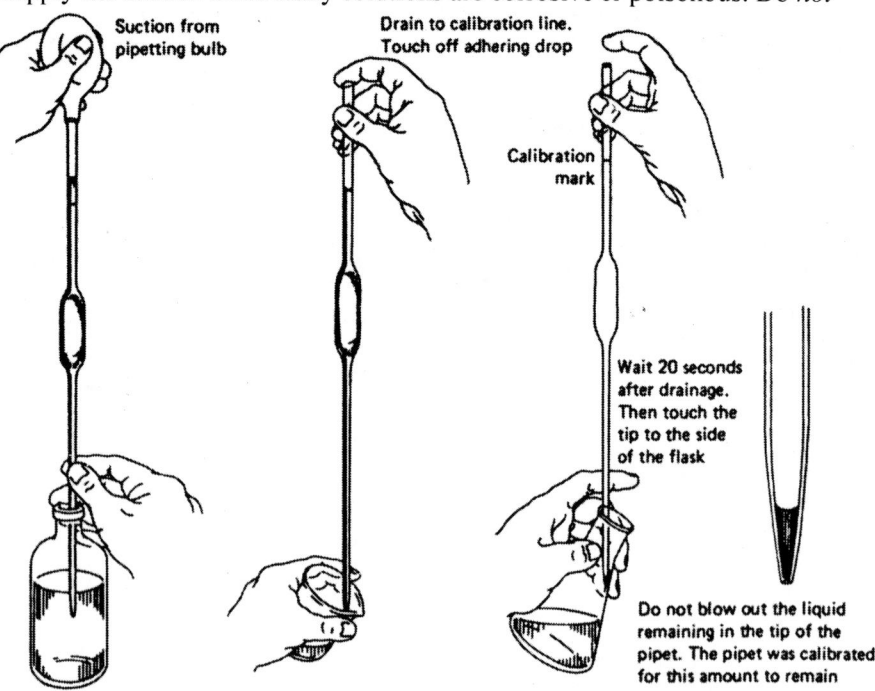

Fig. LT-12. Use of a transfer (volumetric) pipet.

Laboratory Equipment and Techniques

HANDLING AND TRANSFER OF REAGENTS

Careless handling of chemical reagents is dangerous because some chemicals are toxic if swallowed or inhaled, while others may attack the skin, corrode clothing, catch fire, or give off vapors that have unpleasant odors. Prolonged exposure to vapors of any volatile compound is hazardous. Therefore, good technique in handling chemicals is a basic requirement of the laboratory.

Another compelling reason for care in the handling of chemicals is the risk of contaminating a reagent. The results obtained in any experiment depend on the purity of the substances used; use of an impure reagent may lead to erroneous conclusions. Introducing even a very small amount of foreign material into a large container of pure reagent may render it unfit for use. Accidental contamination can be an expensive waste of time and chemicals.

To avoid contamination of reagents, observe the following rules at all times:
1. Read over the entire experiment before attempting any operations. Know in advance what quantities of reagents are required and make sure you have appropriate containers available for them.
2. Take only as much reagent as is required. Once you have removed a portion, return the original container to the storage shelf. Do not return any excess reagent to the original bottle. Share the leftover chemical with another person in the laboratory or dispose of it in the manner recommended by the instructor.
3. Never dip glass tubing or a dropper into a bottle of liquid reagent. When a liquid is to be dispensed in small quantities, its container should have a cap fitted with a dropper. Return the dropper directly to the appropriate bottle after use. The dropper must never be set on a bare desktop or any other surface which might contaminate it. If a reagent bottle is not equipped with a dropper, pour a small quantity of the liquid into a clean container and use your own dropper.

Transferring Liquid Samples

Reagent solutions are often contained in glass-stoppered bottles. When taking a sample from this type of bottle, never allow the part of the stopper that fits into the bottle to become contaminated.

If the stopper has a flat top, either hold the stopper by the top or place the stopper upside down on the table. If the stopper has a vertical flange, grasp the flange between your fingers with your palm turned upward as shown in Fig. LT-13.

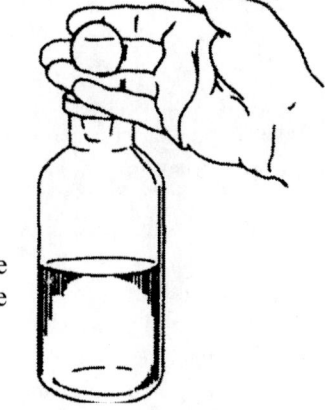

Fig. LT-13. Removing the stopper from a reagent bottle.

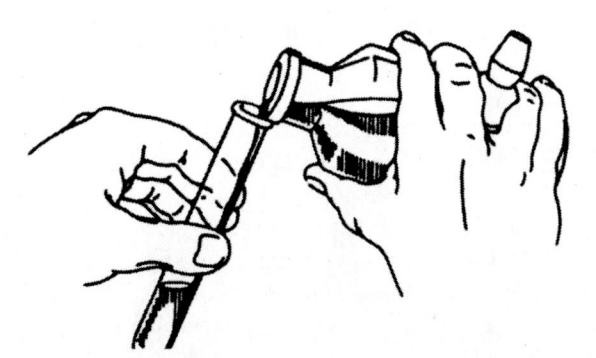

Remove the stopper and then lift the bottle with the same hand, leaving the other hand free to hold the receiving vessel. Bring the neck of the bottle in contact with the rim of the receiving vessel and pour down the side of the vessel so that the liquid does not splash or splatter (Fig. LT-14). Take care that no liquid runs down the outside of the bottle or the outside of the receiving vessel.

Fig. LT-14. Transferring a liquid solution from a reagent bottle to a test tube. One hand holds both bottle and stopper.

Laboratory Equipment and Techniques

To avoid spilling on the outside of the bottle or the receiving vessel, pour the liquid down a stirring rod as shown in Fig. LT-15. If you do spill, carefully wash and wipe the outside of the bottle before returning it to the reagent shelf.

If the reagent bottle has a top that you cannot conveniently hold while pouring from the bottle, set the cap on a clean glass plate or a clean watch glass. To dispense liquid from a beaker, use a stirring rod as an aid in pouring. The stirring rod and beaker can be held with the same hand (Fig. LT-16). Liquids transferred in this way will not dribble down the outside wall of the beaker.

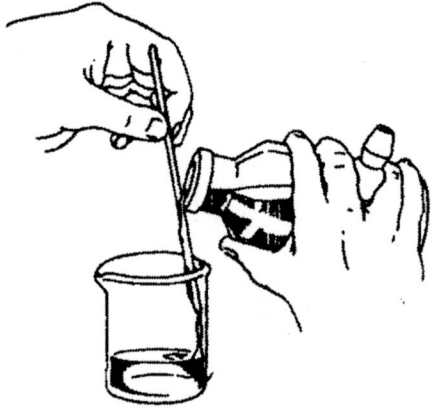

Fig. LT-15. One hand holds both bottle and stopper.

Liquid samples may be weighed in a clean, dry, previously weighed container. If the liquid is volatile, the container should have a suitable stopper, and the stopper must be weighed as part of the container. In subsequent operations, handle the stopper in a manner that avoids contamination of the sample.

Transferring Solid Samples

Remove a solid reagent from the stock bottle by pouring it (gently rotate the bottle) onto a clean watch glass or filter paper. Never insert a spatula or scoop directly into the stock bottle.

Never handle solid chemicals with your fingers - use a spatula or a scoop. Moisture and oils from your skin may contaminate the sample. Also, many chemicals stain, irritate, or harm the skin.

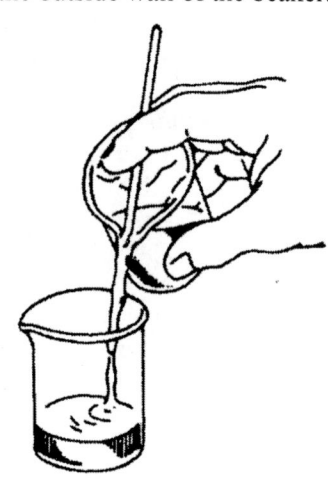

Fig. LT-16. Transferring a liquid solution from a beaker.

Handling Acids and Bases

Concentrated solutions of acids and bases are potentially hazardous because they readily attack skin and clothing. While being mixed with water, concentrated acids and solid bases (such as sodium hydroxide and potassium hydroxide) give off large quantities of heat. If water is poured into a concentrated acid or onto a solid base, the heat liberated may raise the temperature of the water above the boiling point and cause a violent spattering of the corrosive liquid. This could result in badly damaged clothing and injury to the eyes and skin. Therefore, acids and bases should always be added gradually, to water - *never* the reverse!

When using the common acids and bases provided on the laboratory shelf, pour an approximate quantity into a clean beaker or a clean graduated cylinder. Stopper the original bottle and return it to the shelf. Measure the exact amount needed into a graduated cylinder. To dispose of any excess acid or base, slowly add the liquid to a large quantity of water and then dilute it with more water. Pour the diluted mixture into the sink and wash it down the drain with lots of water. Should some acid be spilled, wash it off first with copious amounts of water and then with a dilute solution of sodium bicarbonate. A spilled base should be similarly diluted with water, wiped up, and the area washed with dilute acetic acid and rinsed with water.

Laboratory Equipment and Techniques

METHODS OF FILTRATION

Gravity Filtration

The separation of solids from liquids can be accomplished through gravity filtration. A funnel, filter paper, and a ring stand are used to make the gravity filter (Fig. LT-17). Before filtration, allow solid particles in the solution to settle. Then pour into the funnel small portions

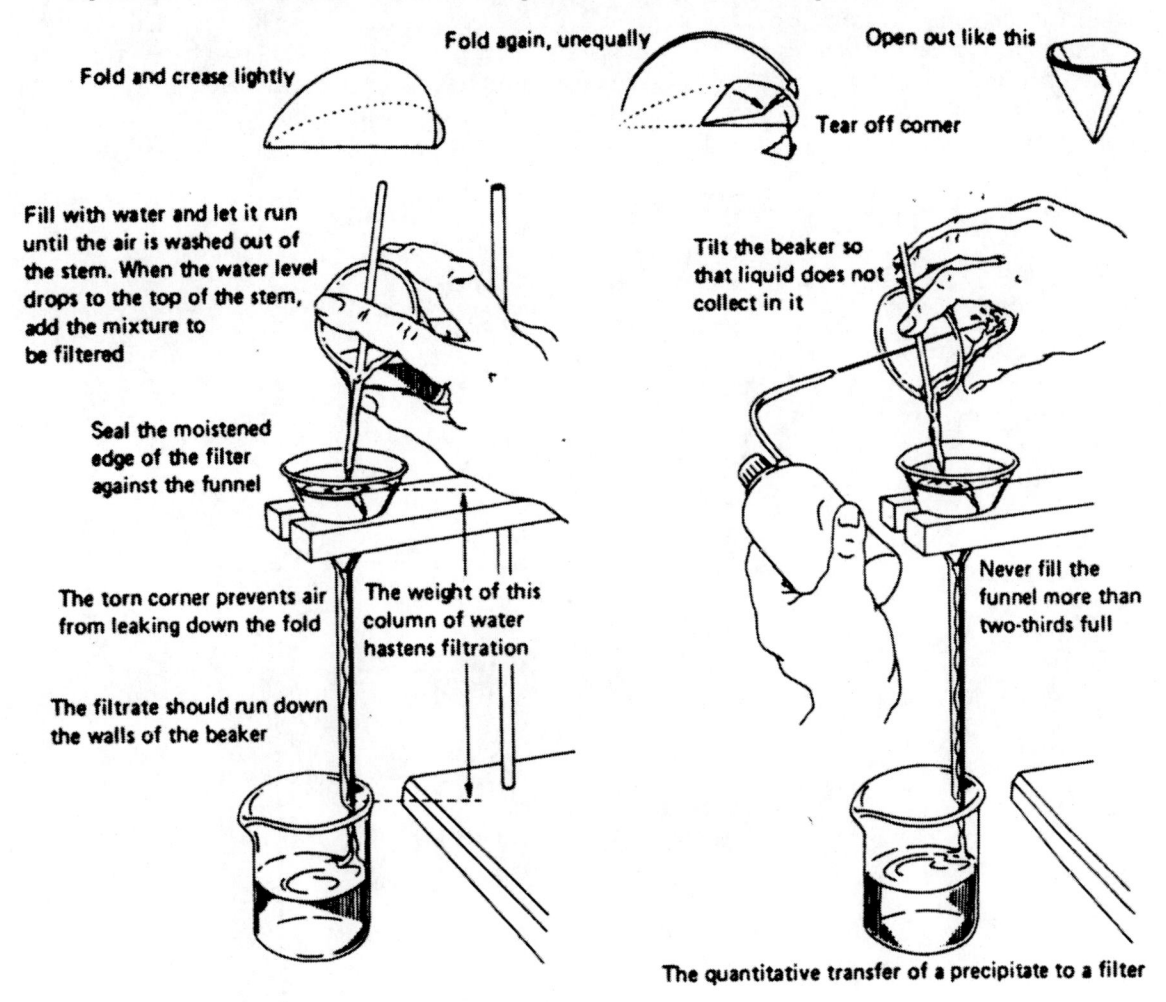

Fig. LT-17. Techniques of gravity filtration

of the liquid (called the supernate) that rises above the solid sediment. The technique of pouring off the supernate is called decantation. The funnel should never be more than two-thirds full of liquid at any time. Next, use a plastic wash bottle and a stirring rod to wash the solid sediment into the funnel (Fig. LT-17). The clear solution that passes through the filter paper is called the filtrate. The residue that remains on the filter paper is called the precipitate.

Buchner Filtration

A more rapid method of separation of liquids from solids can be achieved through Buchner filtration (Fig. LT-18). The filter consists of a Buchner funnel, filter paper, a ring stand, and a vacuum source. Appropriate clamps must be used to hold the filter flask in place. Also, a safety trap is used in conjunction with Buchner filtration to prevent any liquid in the vacuum source from contaminating the filtrate, or vice versa. A filter paper slightly smaller than the circular perforated plate is centered in the Buchner filter funnel and moistened with a few drops of water. The filter paper seals to the plate when a light vacuum is applied.

The procedure for transferring material to the filter from a container is the same as for gravity filtration. First, allow solid particles to settle, and then decant the supernate into the funnel while maintaining gentle suction on the filter. Wash most of the solid sediment and any remaining liquid into the Buchner funnel with a wash bottle.

Increase the vacuum only when sufficient solid has accumulated on the paper to strengthen it. (A strong vacuum applied to moist paper may cause it to tear.) To wash the solid in the funnel, distribute small amounts of wash liquid over the surface and allow the liquid to be sucked through the precipitate into the filter flask. To dry the solid, keep it in the funnel as the vacuum draws air through the precipitate. You can remove the precipitate by disconnecting the vacuum, then inverting the funnel over a clean piece of paper and tapping it. Lifting an edge of the filter paper with a spatula helps dislodge the filter.

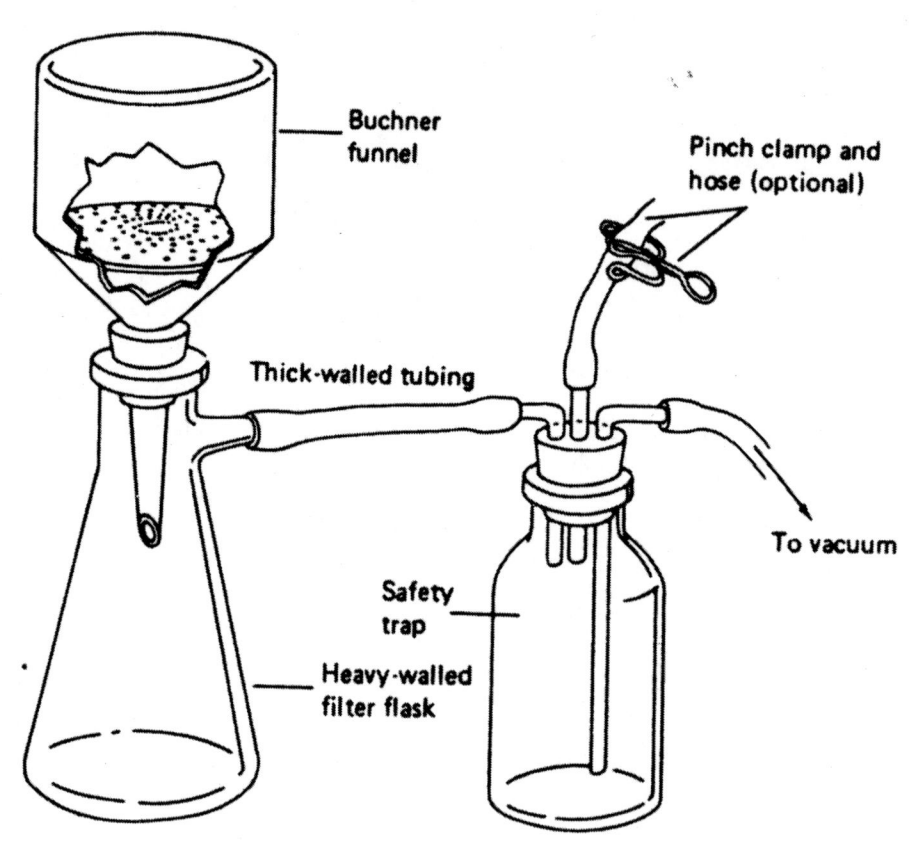

Fig. LT-18. Buchner filtration (appropriate clamps omitted to improve clarity.)

Mathematical Treatment of Data

DIMENSIONAL ANALYSIS

Calculations made in the chemistry laboratory are concerned with real objects and quantities. Therefore, the numbers involved represent the values of specific measurements. The value reported for a measurement can have meaning only if the units of the measurement are stated along with its numerical magnitude. Scientific measurements are expressed in units of the metric system. An expanded and improved version of the metric system is called the Systeme International d'Unites, or SI. SI units are described in Appendix B, Tables 9-11 and 15.

Each measurement included in any calculations, and the final result, *must be expressed in appropriate units*. For example, if an object has a mass of 36 g and a volume of 24 mL, its density is calculated to be

$$d = \frac{m}{V} = \frac{36 \text{ g}}{24 \text{ mL}} = \frac{1.5 \text{ g}}{\text{mL}} = 1.5 \text{ g mL}^{-1}$$

Grams per milliliter, abbreviated g/mL or g mL^{-1}, is an appropriate combination of units for density. Often it is desirable to convert a measurement made in one set of units to another set of units. A general procedure for making such conversions is known as the factor-label method. In the factor-label method, units and dimensions are treated like algebraic variables and are combined or canceled in such a manner that the correct units for the result are obtained. To convert a measured quantity to a different set of units, the quantity is multiplied by a factor that is the ratio of equivalent quantities in the two sets of units. For example, 1.000 lb in the English system of measurement is equivalent to 453.6 g in metric units. Therefore, the ratios

$$\frac{1.000 \text{ lb}}{453.6 \text{ g}} \quad \text{and} \quad \frac{453.6 \text{ g}}{1.000 \text{ lb}}$$

are appropriate for converting grams to pounds and pounds to grams. In the following example the mass 3,744 g is converted to pounds.

$$3{,}744 \text{ g} \times \frac{1.000 \text{ lb}}{453.6 \text{ g}} = 8.254 \text{ lb}$$

Note that the factor is chosen so that the unit in the denominator of the ratio cancels the unit of the given quantity; thus, the final result has the proper units. Indeed, if there is a question of how to undertake a particular calculation, consideration of the units will suggest a proper procedure.

For example, given that the specific heat of water is 4.18 J g^{-1} deg^{-1} how many joules (J) are required to raise the temperature of 27 g of water by 30°C? The desired result is a number of joules. The data given include a mass (grams) of substance (water), a property of water (its specific heat in joules per gram degree), and a temperature interval (in degrees). When these units are combined algebraically by multiplication, the result is in joules. (Note that units cancel whether they are singular or plural.)

$$\frac{\text{gram} \times \text{joules} \times \text{degrees}}{\text{gram} \cdot \text{degrees}} = \text{joules}$$

Thus, the suggested procedure is to multiply the mass of water by its specific heat and by the indicated temperature change.

$$27 \text{ g} \times \frac{4.18 \text{ J}}{\text{g} \cdot \text{deg}} \times 30 \text{ deg} = 3.4 \times 10^3 \text{ J}$$

(See the *Significant Digits* section below to explain rounding off in this example.)

As the example shows, when using the factor-label method, often you need not make actual numerical calculations until the final step. This practice allows for the most efficient use of a calculator or a computer. Moreover, setting up a calculation in terms of units and saving the numerical operations until the final step makes it easier to check for inadvertent mistakes.

Regardless of the method of computation, the final result of any calculation must be expressed to the proper number of significant digits. A discussion of significant digits follows.

UNCERTAINTY AND ERROR ANALYSIS
Significant Digits

Any measurement has an uncertainty that depends upon the reliability of the measuring device. Therefore, numbers that represent measurements are approximations rather than exact quantities. The concept of significant digits is used to indicate the magnitude of these approximations. The significant digits in a number are those digits that are known with certainty plus one estimated digit. For example, a thermometer which is graduated every 0.5°C indicates a temperature slightly less than halfway between 35°C and 35.5°C. The temperature is recorded as 35.2°C. Three significant digits are used: 35 is known with certainty, but the 0.2 is estimated.

1. A zero is significant only when it is used to indicate the accuracy of measurement. For example, 5.000 has four significant digits. Zeros used simply to locate the decimal point are not significant. Thus, 0.0025 has only two significant digits.

   With a number such as 50,000 it is not possible to tell whether the number represents an estimate or an exact number. Are there five significant digits (implying precisely 50,000) or are there fewer significant digits (implying an estimate of 50,000)? To avoid this difficulty such numbers should be written in exponential notation, where the significant digits are indicated in the pre-exponential factor. Hence, the number 5.000×10^4 represents 50,000 with an uncertainty of 5. The number 5×10^4 represents an estimate of 50,000 with an uncertainty of 5,000.

2. Exact numbers which are *not* measured values, such as the 1,000 mL in 1 L, are considered to have an infinite number of significant digits. (That there are 1,000 mL in 1 L is a definition, not a measurement.) If a calculation requires the conversion of a measured volume from liters to milliliters, the number of significant digits in the result is determined only by the uncertainty in the measured volume.

Calculations Using Significant Digits

If a calculated numerical answer contains more significant digits than are justified by the results of an experiment, the numerical answer must be rounded off to the correct number of significant digits.

1. If the leftmost digit to be dropped is more than 5, the last remaining digit is increased by one: 4.962 rounded off to two significant digits becomes 5.0.

2. If the leftmost digit to be dropped is less than 5, the last remaining digit stays the same: 4.948 rounded off to two significant digits becomes 4.9.

3. If the leftmost digit to be dropped is 5, with no additional digits to the right other than zero, the last remaining digit is rounded off to the nearest even number. When rounded off to two significant digits, 4.55 becomes 4.6, 4.850 becomes 4.8, but 4.859 becomes 4.9.

4. For addition and subtraction, the last digit retained in the answer should correspond to the first doubtful decimal place in any of the added numbers.

7.49	856.8	956.6
+0.4246	+ 1.327	- 0.0034
7.9146 → 7.91	858.127 → 858.1	956.5966 → 956.6

5. In multiplication and division, the answers should contain no more digits than the least number of significant digits in any of the numbers entering the calculation.

$$\frac{0.356}{3.8} = 9.4 \times 10^{-2}$$

$$(1.8 \times 10^1) \times 3.14 \times (7.003 \times 10^{-6}) = 4.0 \times 10^{-4}$$

6. In logarithms, the characteristic (in the case below, the number 1, the number to the left of decimal point) does not count as a significant digit. The logarithm of $44.4 = \log (4.44 \times 10^1) = \log 4.44 + \log 10^1 = 0.647 + 1 = 1.647$. In general, the number of significant digits in the mantissa (in this case 0.647, the number to the right of the decimal point) should reflect the number of significant digits in the original number.

Precision and Accuracy

The difference between an observed value and the "true" value is known as the error of the measurement. The reported value for a physical measurement is of little value unless it is accompanied by an indication of its accuracy.

The *accuracy* of a measurement describes the difference between an observed value and the "true" value of the quantity being measured. In some cases, the "true" value is given by a definition, but in many cases "true" values are not known. If several reliable investigators independently obtain the same value for a measured quantity, then that value is assumed to be the "true" value.

Precision is concerned with the reproducibility of a result by the same method of measurement. If the difference in the numerical results of a series of measurements is small, the measurements have a high precision. Conversely, if the difference is large, the measurements have a low precision. Precision is not the same as accuracy. A defective instrument may give an inaccurate result with a high precision. Consider the following series of measurements, made on three different digital balances, of the mass of a metal bar having a standard or "true" mass of 25.00 g.

Balance A	*Balance B*	*Balance C*
25.01 g	25.07 g	25.00 g
25.02 g	25.06 g	24.92 g
24.99 g	25.08 g	25.13 g
24.98 g	25.05 g	25.19 g

For balance A, the average of the measurements is 25.00 g; for balance B, 25.06 g; and for balance C, 25.06 g. The accuracy and precision of the three balances are shown in Fig. M-1.

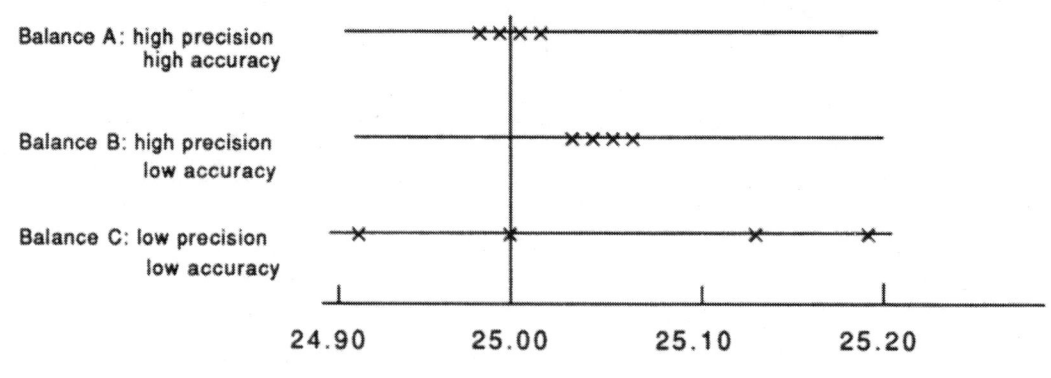

```
Balance A: high precision     ××|××
           high accuracy
                          ────────────────────────────

Balance B: high precision              ××××
           low accuracy
                          ────────────────────────────

Balance C: low precision  ──×─────|─────────×──────×──
           low accuracy
                          └──────┴─────┴─────┴──────┴
                        24.90   25.00  25.10   25.20
```

Fig. M-1. Accuracy and precision.

Measurements made on balance B and on balance C would appear to be equally valid if only the average values are considered. However, because of the lack of precision in measurements made with balance C, reported values should be expressed to only three significant digits (25.1 rather than 25.13). Measurements made with balance A and balance B would be reported to four significant digits. Because the mass of the standard is known in advance, the results obtained on balance A should be preferred. The accuracy of a given instrument can be ascertained only by comparison with an acceptable standard. The process of comparison with a standard is called *calibration*.

The precision of various common laboratory instruments is given in the following list

Instrument	*Precision*
digital balance	0.01 g
50-mL buret	0.02 mL
50-mL graduated cylinder	0.2 mL
10-mL graduated cylinder	0.1 mL
110°C thermometer	0.20°C
25-mL transfer (volumetric) pipet	0.02 mL
10-mL transfer (volumetric) pipet	0.01 mL

Error

Experimental measurements are subject to error resulting from experimental technique, faulty instruments, or poorly calibrated equipment. Error of this kind is called *systematic error*. A systematic error causes each measurement to be in error in the same direction. It reduces the accuracy without necessarily reducing the precision of a measurement. Examples of apparatus that exhibit systematic error are a bathroom scale that always reads 5 lb. too low and a wristwatch that is set 5 minutes too fast. Systematic errors can be eliminated by calibration and by careful control of experimental conditions.

However, despite the experimenter's best efforts to remove all systematic error from any system, several readings made with the same instrument on the same object will vary about some average value by small random amounts. Measurements in which errors of this type are found are said to be subject to *random error*.

For example, the mass of a coin measured many times on a digital beam balance will usually show an uncertainty of ± 0.01 g. The construction of the balance prevents obtaining a more precise value for the mass of the coin.

Mathematical Treatment of Data

Frequency of Random Error

When a large number of measurements $(x_1, x_2, x_3 ...)$ are made on an object or a physical quantity, x, they may be subject to random error. For convenience, let x_i denote the value of an arbitrary measurement, and let T be the "true" value of the quantity x. The difference between the value of x_i and the "true" value, T, is the random error of the measurement. The distribution of random errors about T may take the form of a normal error distribution as shown in Figs. M-2 and M-3.

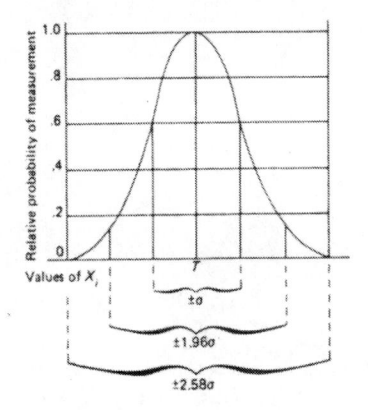

Fig. M-2. The normal distribution curve.

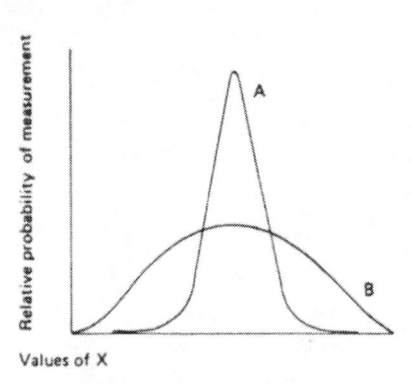

Fig. M-3. Measurements of the same quantity with different levels of precision. Curve A indicates high precision; curve B, low precision.

The spread or dispersion of measured values about T is defined by σ, the standard deviation. The magnitude of σ is such that approximately 68% of all measurements lie within $\pm \sigma$; 95% within $\pm 1.96\,\sigma$; and 99% within $\pm 2.58\,\sigma$. Thus, the smaller the value of σ, the more precise the measurement. The normal error curve is based on graphing the results of an infinite number of measurements. However, experiments are normally conducted on only N measurements, where N is less than 10. Nevertheless, it is possible to find the precision of your results regardless of the number of measurements. Generally, the mean or average value of the set of measurements is the best approximation to the "true" value. The difference between each individual measurement, x_i and the mean value, $\bar{x}$ is called the deviation δ_i. Hence,

$$\delta_i = \bar{x} - x_i.$$

The deviation and the error are identical when the mean value exactly equals the "true" value. An accurate measure of precision is related to the variance, S^2, which is defined by

$$S^2 = \frac{\Sigma \delta_i^2}{(N-1)}$$

The symbol Σ stands for the summation of all the δ_i^2. It is important to determine the variance for a series of N measurements because, for a normal error distribution, the square root, S, of the variance is the standard deviation. Therefore, the probability that a measurement lies within $x \pm S$ is 68%, $x \pm 1.96\,S$ is 95%, and $x \pm 2.58\,S$ is 99%.

Calculation of the variance is straightforward and is easily performed with the aid of a pocket calculator. First, the average ($\bar{x}$) of all N measurements is calculated. Then the deviation (δ_i) and the square of the deviation (δ_i^2) are calculated for each measurement. The sum of all the squares of the deviation ($\Sigma \delta_i^2$) is divided by $N - 1$. Consider the following example for $N = 5$ measured values (x_i).

Mathematical Treatment of Data

x_i	$\delta_i = \bar{x} - x_i$	δ_i^2
2.57	-0.01	0.0001
2.58	-0.02	0.0004
2.54	+0.02	0.0004
2.52	+0.04	0.0016
2.61	-0.05	0.0025

The last column sum ($= \Sigma \delta_i^2$) = 0.0050

The average or mean value = $\bar{x}$ = 2.56

$$\frac{\Sigma \delta_i^2}{N-1} = \frac{(0.0050)}{4} = 0.00125 = S^2$$

$$S = 0.035$$

The 95% Confidence Level

Often it is interesting to know within what range 95% of all experimental observations are expected to fall. This range is also known as the 95% *confidence level*. Statistical methods enable its easy calculation. For an infinite number of measurements, it can be shown that the mean, $\bar{x}$, becomes identical with T, the "true" value. In such a case, the 95% confidence level is defined by $\pm 1.96S$. However, if there are only a small number of measurements from which to evaluate x, the mean may differ significantly from the "true" value. Furthermore, the value of S calculated using the mean may be different from the value of S associated with the "true" value. To allow for these possibilities, there is a correction factor denoted as t for each value of N, the number of measurements.

$N-1$	t
1	12.7
2	4.30
3	3.18
4	2.78
5	2.57
6	2.45
7	2.36
8	2.31
9	2.26
10	2.23
15	2.13
300	2.04
∞	1.96

Table M-1. Critical values of t, 95% confidence value

The 95% confidence level is calculated from the equation:

$$95\% \text{ confidence} = \frac{tS}{(N^{1/2})}$$

The value of t is taken from Table M-1, corresponding to the value of $N-1$. More complete values of t can be found in the *Handbook of Chemistry and Physics*. Using the values from the example on variance,

$$95\% \text{ confidence} = \frac{(2.78)(0.035)}{5^{1/2}} = 0.044$$

The best value of the measurements is written as 2.56 ± 0.04 (95% confidence, $N = 5$).

COMMENTS

Mathematical Treatment of Data

Experiment 1
Graphing Data with Excel®

OBJECTIVES

To acquire graphing skills, to express physical properties using mathematical relationships; and to communicate physical properties using graphical methods.

EQUIPMENT

Personal computer running Microsoft Excel® 2010 or better.

INTRODUCTION[1,2,3]

Graphing is a technique of presenting raw or converted data into a form that it may be interpreted. A graph can be linear, exponential, non-linear or non-exponential in nature depending on the relationship between the quantities being graphed.

Graphing shows relationships between two or more quantifiable values. Graphing may also be used to illustrate trends when comparing two or more quantifiable values.

Consider the following real-life applications of graphing:

1. Illustrating stock market patterns for a particular country or for the whole world over a defined period of time.
2. Showing the rate of decay of a radioactive substance over a defined period of time.
3. Presenting the number of patients admitted to a hospital in a day, a week, a month, or a year. Similarly, if a hospital wanted to track the number of patients that came in with influenza versus their demographic background, a bar graph could be constructed to illustrate that data (Fig. 1-1).

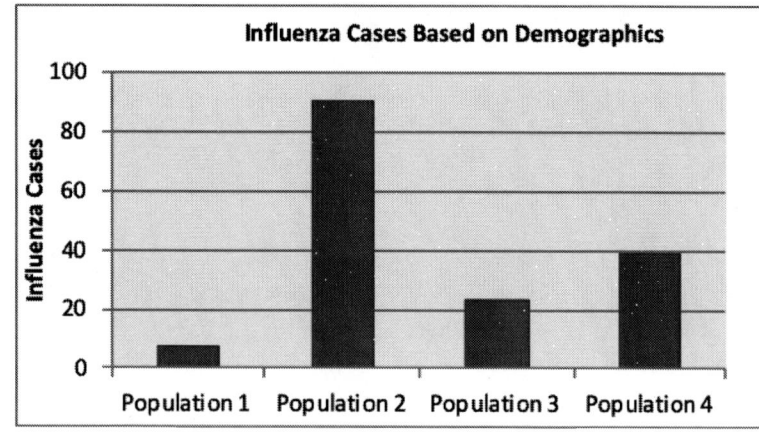

Fig. 1-1. Example of a bar graph

Graphing in Excel® is simple, if you follow certain rules.

- X-data must be in a column at the far left.
- Y-data should be in columns to the right of the X-data.

Generally, the X-axis (horizontal axis) contains the value YOU MANIPULATED (independent variable) and the Y-axis (vertical axis) contains the measurements taken (dependent variable).

Example: You poured a specific volume of an unknown liquid several times consecutively into a graduated cylinder and measured the mass for after each addition. Volume is the independent axis (X) and mass, is the dependent (Y) axis.

PROCEDURE

The following data (Table 1-1) were recorded in an experiment to determine a linear relationship between mass (g) and volume (mL) of a new organic solvent (liquid) produced.

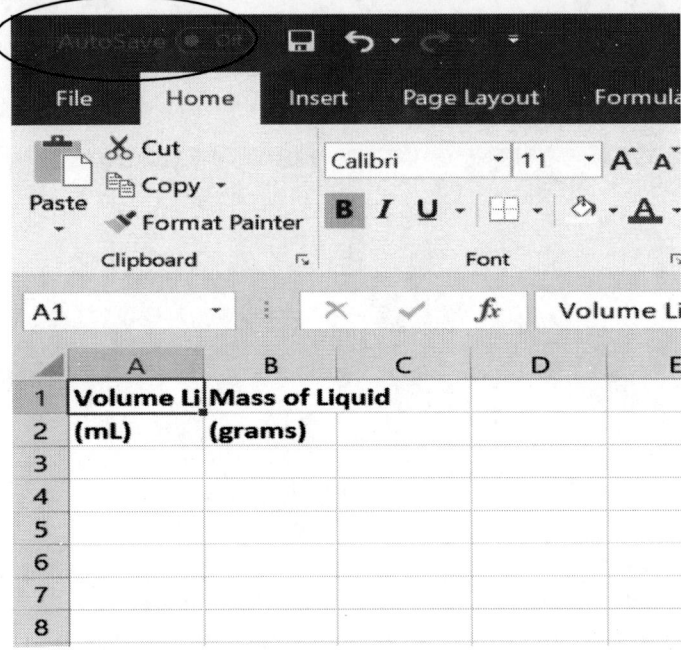

Volume Liquid (mL)	Mass of Liquid (grams)
23.0	26.8
34.0	38.0
45.0	53.6
59.0	70.5
68.0	80.3

Table 1-1. Example bar graph

Fig. 1-2. Excel® data entry

1. Type this data into your Excel® file. In the above example the "X" data (independent variable) is the volume and should be typed into a column on the far left.
2. The "Y" data (dependent variable) is the mass of the liquid and should be typed into the column adjacent to the volume column (Fig. 1-2).
3. Highlight the data that you wish to include on your chart (graph).
4. Click on Insert (chart menu is displayed) on the Toolbar and then click on scatter (Fig. 1-3). For this exercise choose the first option (a series of dots with no lines.)

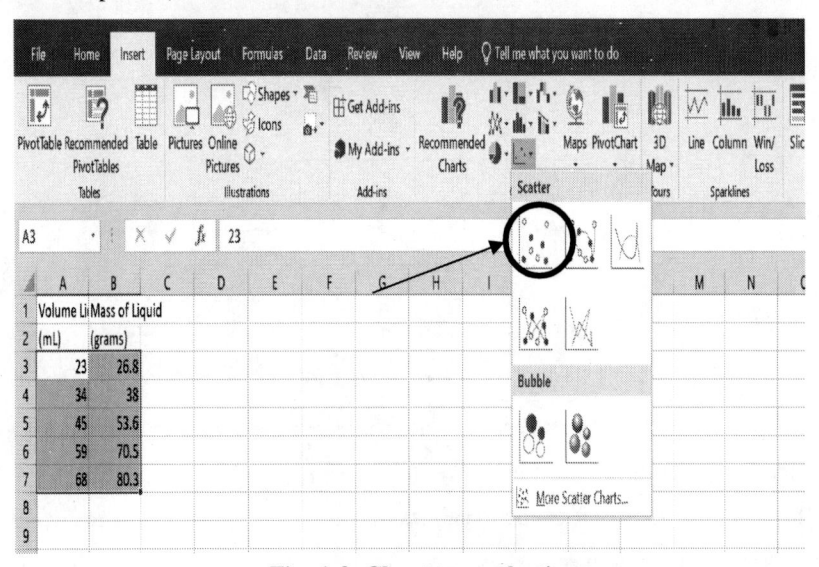

Fig. 1-3. Chart type selection

Experiment 1 Graphing Data with Excel®

5. Under "Design" click the "plus sign" which will show as chart elements. At the "chart elements", you have the option to add a trendline, a chart title, and label the X and Y axes. To obtain the equation on chart, display R-Square value on chart and confirm that the graph is linear. Select "more options" under the trendline icon showing below (Fig. 1-4) and

   a. Select "Display equation on chart"
   b. Select "Display R-squared value on chart"

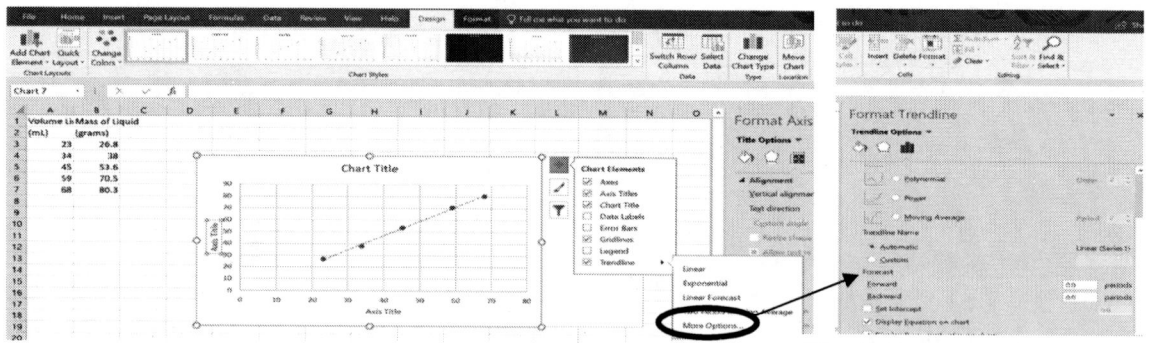

Fig. 1-4. Trendline, display equation and R-squared

6. Enter a **title:**

   - While still in the design screen, under chart elements click "Chart Title" and type the title inside the chart title displayed here (Fig 1-4). The title should have the format "Mass of Liquid vs Volume".

7. Enter **x-axis and y-axis titles** WITH UNITS! With only one set of data on a graph, you will not need a legend. Legends are used to identify each data set, and are only useful when there are several different sets of data on one graph.

8. Design-Axis Title-Primary Horizontal Axis Title (or Primary Vertical Axis Title) – Selection position of title.

Example

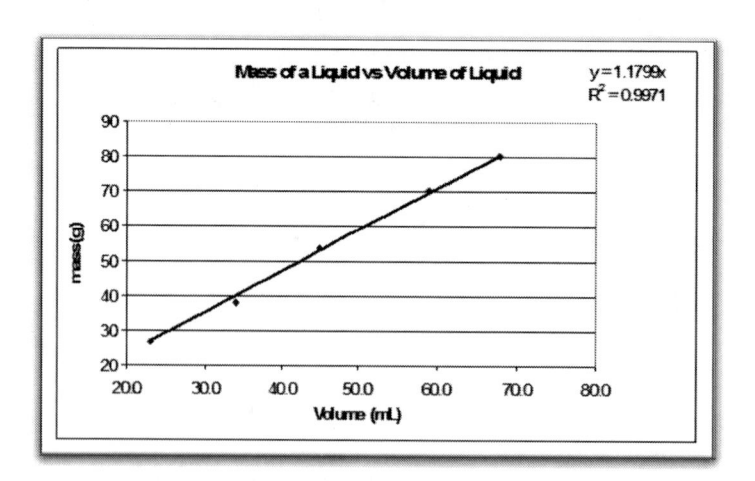

What are the **units** for the slope in this example?

What is the significance of the slope in this example (*i.e.,* what physical property is determined)?

Write the equation for this line.

Experiment 1 Graphing Data with Excel®

FURTHER READING
1. Tufte, Edward R (2001), *The Visual Display of Quantitative Information* (2nd ed.), Cheshire, CT: Graphics Press
2. Billo, E. J. (2011). *Excel for Chemists, 3rd ed.*, John Wiley & Sons
3. Verschuuren, G; *Excel 2013 for Scientists,* https://www.ipgbook.com/excel-2013-for-scientists-products-9781615470259.php (last accessed July, 2019)

COMMENTS

Name_____ Lab Section_____ Date_____

Prelaboratory Assignment: Experiment 1
Graphing Data with Excel®
Where appropriate, answers should be given to the correct number of significant digits.

1. What is an independent variable?

2. What is a dependent variable?

3. Give three (3) examples for each question (1) and (2) above.

4. Where can you apply this information in a real-life situation?

5. A physical science student carried out an experiment to determine the mass – volume relationship in hydrogen gas at standard temperature and pressure. The following data were collected as presented in the table. Use the data to derive a relation between mass and volume. What relationship exists between the two physical properties, what is the unit of the slope? What is the name of the slope? The following data were recorded.

Mass (mg)	Volume (mL)
33.0	368
44.0	480
55.0	636
69.0	805
78.0	903

6. Several years ago, in Japan, a natural disaster occurred where nuclear reactors exploded with the destruction of property and the loss of many lives. As a chemist how will you determine the rate that the radioactive elements released by these reactors will decay in soils in the area? Suggest a type of relationship that can be used to study this problem in order to advise inhabitants about the safety for return settlements. Support your findings with a graph.

Experiment 1 Graphing Data with Excel®

Name_____ Lab Section _____ Date_____   43

REPORT ON EXPERIMENT 1
Graphing Data with Excel®

Where appropriate, answers should be given to the correct number of significant digits.

QUESTIONS
(Submit your answers on a separate sheet as necessary.)

1. The following graph was constructed from mass and volume data for aluminum and for copper, respectively. Use the graph to answer the following questions.

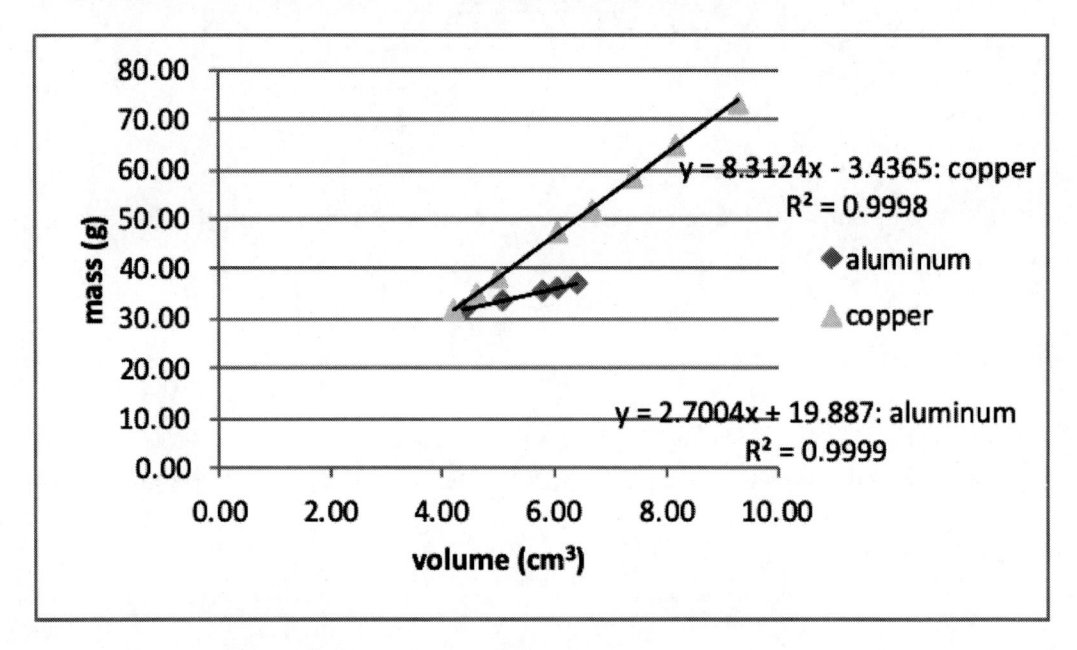

   [Graph: mass (g) vs volume (cm³). Copper: y = 8.3124x − 3.4365, R² = 0.9998. Aluminum: y = 2.7004x + 19.887, R² = 0.9999.]

   a. What is the mass of 5.00 cm³ of copper?

   b. According to this data what is the density of aluminum?

   c. If you were consulted to create a new space jet to travel to Jupiter, which of these metals would you recommend to be used in the building of the jet? Give three reasons to support your answer.

2. An experiment was conducted in which the pressure of a gas was measured at different temperatures at constant volume. The results are tabulated below.

Pressure (psi)	14.6	15.9	17.6	19.5
Temp (K)	273	305	325	362

   a) If you were to graph this data, which is the dependent variable? Explain your choice.

Experiment 1 Graphing Data with Excel®

3. An experiment carried out by a forensic scientist determined the density of an unknown liquid found at a crime scene. Below is the data for the unknown liquid.

Mass (mg)	Volume (mL)
15	12
23	16
35	22
46	24
56	27
70	35

   a. With your knowledge about Excel®, create a suitable graph for the data above (don't forget a Title for your graph.)

   b. What kind of relationship exists between the two parameters measured above? Using your graph, support your answer with mathematical evidence.

   c. If the mass of the last entry in the table is doubled, what will be the volume of the unknown liquid?

Experiment 1 Graphing Data with Excel®

Experiment 2
Identification of Compounds through Physical and Chemical Changes

OBJECTIVE

To identify compounds on the basis of physical and chemical changes and design a procedure for determining the presence of a particular compound.

EQUIPMENT

Fifteen 12 x 75-mm test tubes, test tube rack, test tube holder, permanent marker (to label test tubes), three medicine droppers. If test tubes are not available, a 24-well plate may be used.

REAGENTS

Part A: Approximately 1 mL (~5-10 drops) each of aqueous known solutions of NaCl, Na_2CO_3, $MgSO_4$, NH_4Cl, and H_2O (this step is repeated for the three test reagents); 5-10 drops each of aqueous test reagents of $AgNO_3$, NaOH, and HCl.

Part B: Approximately 5-10 drops each of solutions labeled 1 – 5; 5-10 drops of test reagents labeled A, B, and C.

SAFETY AND DISPOSAL

Refer to the MSDS information available online when working with NaCl,[1] Na_2CO_3,[2] $MgSO_4$,[3] NH_4Cl,[4] $AgNO_3$,[5] NaOH,[6] and HCl.[7] Contact with silver nitrate solutions can lead to staining of the skin, dermatitis, and painful burns.[5]

Disposal should be in accordance with local, state and federal regulations. Disposal for NaCl, Na_2CO_3, $MgSO_4$, NaOH, and HCl should be into a labeled laboratory waste container for inorganic chemicals. All waste silver salts, especially those that result from the $AgNO_3$ reagent observations found in this experiment, should be recovered in a specially labeled waste container for silver salts.

INTRODUCTION

A goal of a research chemist is, for example, to separate the substances of a reaction mixture and attempt to identify each substance through a systematic study of their chemical and physical properties.[1] Any change in the appearance of a substance can be classified as either a *physical* or *chemical* change.

In a *physical* change, the same substance remains after the change (melting or boiling a liquid, cutting a piece of wood, tearing paper, dissolving sugar in water, and pouring a liquid from one container to another are some examples). These changes do not change the chemical character of the substance.

In a *chemical* change, the substance undergoes a change so that one or more new substances with different characteristics are formed (burning, digestion, and fermenting are some examples). Let us take a specific example. Sodium is a silvery, soft metal that reacts vigorously with water. Chlorine is a yellow-green gas that is highly corrosive and poisonous. However, if these two elements are combined, they produce a different substance, a white crystalline solid. This new substance is common table salt, sodium chloride, which is neither reactive with water nor poisonous!

As a rule of thumb, if there is gas formation, color change, precipitation, a reaction that generates or removes energy, or odor change occurs, a *chemical* change has taken place.

A *gas* forms. This evolution may be quite rapid, or it may be a "fizzing" sound. For example, the bubbles of gas observed when antacid is dropped into water.

A *color change* occurs. A substance added to the system may cause a color change. In describing a colored liquid, name the color. In describing a liquid that has no color, call it colorless. For example, milk would be described as a cloudy white liquid. If it is a transparent, colorless liquid, describe the liquid as clear and colorless.

A *precipitate* appears (or disappears). The nature of the precipitate is important. It may be crystalline; it may have color or it may merely cloud the solution. An example is shown in Fig. 2-1.

An *energy change* (heat or absence of heat) occurs. Heat may be evolved or absorbed. The reaction vessel becomes warm if it releases heat (exothermic) or cools if it absorbs heat (endothermic).

An *odor change* is detected. The odor of a substance may appear, disappear, or become more intense.

Part A of this experiment asks you to observe chemical reactions of various compounds and identify these compounds based on their chemical properties.

Fig. 2-1. Precipitate formation

In Part B you will be given an unknown compound (one that you have previously investigated in Part A) to identify. The interpretations of the collected data (from Part A) will help you in identifying your unknown.

The chemical properties of the following **known** compounds, dissolved in water, will be investigated in Part A:

sodium chloride	$NaCl$ (*aq*)	ammonium chloride	NH_4Cl (*aq*)
sodium carbonate	Na_2CO_3 (*aq*)	water	H_2O (*l*)
magnesium sulfate	$MgSO_4$ (*aq*)		

The following **reagents** will be used to identify and characterize the above compounds:
- silver nitrate $AgNO_3$ (*aq*)
- sodium hydroxide $NaOH$ (*aq*)
- hydrochloric acid HCl (*aq*)

Note: *You should discuss your findings and interpretations with a partner, but each of you should analyze your own unknown. Remember, you will be conducting a test on each known compound and your unknown with a single test reagent. To organize your work, there is a "reaction matrix" provided for you to describe your observations. Because the space is limited, you may want to use the following codes:*

  pc – precipitate + color
  cc – cloudy + color
  nr – no reaction
  g – gas, no odor
  go – gas, odor

Experiment 2 Physical and Chemical Changes

PROCEDURE
Wear your safety goggles at all times in the laboratory.

Part A

1. *Silver nitrate (AgNO₃) reagent observations.*
   Label five small, clean test tubes with a permanent marker (Fig. 2-2). Alternatively, you can use a clean 24-well plate (Fig. 2-3). You should ask your lab instructor which setup to use.

   Place 5-10 drops of each of the five "known" solutions into your labeled test tubes (or wells A1-A5, Fig. 2-3).

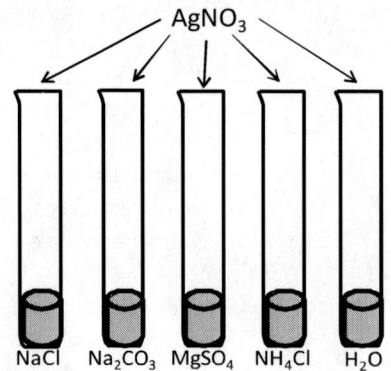

   Use a dropper bottle (or a dropper pipet) to deliver 3-5 drops of $AgNO_3$ solution. (**Caution**: *$AgNO_3$ forms black stains on skin that can be painful. Thoroughly rinse your hands after working with silver salts.*) If you observe a chemical change, add 5-10 more drops to see if there are any additional changes. Record your observations in the "reaction matrix".
   SAVE YOUR TEST SOLUTIONS for part A. 4 of your report. You will need to write the formula for each precipitate that forms. For example, as shown in Fig. 2-1, a precipitate of AgCl(*s*) will form when you mix solutions of NaCl(*aq*) and $AgNO_3$(*aq*). You may ask your lab instructor for assistance. You may also refer to a solubility table if needed.

   Fig. 2-2. Arrangement of test tubes for $AgNO_3$ test

2. *Sodium hydroxide (NaOH) reagent observations.*
   Follow the same procedure as in part A.1.a. above for your second set of test tubes (or wells B1-B5, Fig. 2-3).

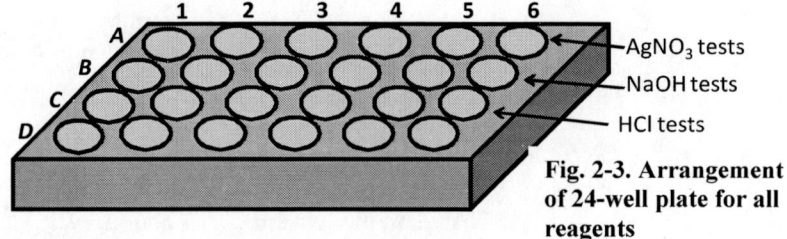

   Fig. 2-3. Arrangement of 24-well plate for all reagents

   To each test tube, slowly add 5-10 drops of NaOH; make observations as you add the NaOH. In particular, observe if any gas evolves in any of the tests. Check for odor.

   SAVE YOUR TEST SOLUTIONS for part A. 4 of your report. You will need to write the formula for each precipitate that forms.

3. *Hydrochloric acid (HCl) reagent observations.*
   Follow the same procedure as above for your third set of test tubes (or wells C1-C5, Fig. 2-3). To each test tube, slowly add 5-10 drops of HCl; make observations as you did for the NaOH. In particular, observe if any gas evolves in any of the tests. Check for odor.
   SAVE YOUR TEST SOLUTIONS for part A. 4 of your report. You will need to write the formula for each precipitate that forms.

4. *Identification of unknown.*
   Obtain an unknown for Part A from your instructor. Perform the three reagent tests on your unknown. Identify the unknown compound based on the data from the 'known' solutions collected in your "reaction matrix".

Experiment 2 Physical and Chemical Changes

Part B

The Part B experiment design is similar to that of Part A. Thus, you will need either 15 clean test tubes or a clean 24-well plate.

1. *Prepare solutions*

    Place ~1 mL of each test solution (there are five solutions labeled 1 through 5) in the clean test tubes or in the well plate.

2. *Test solutions*

    Test each of the five solutions with reagent A. If, after adding several drops, you observe a chemical change, add 5-10 drops more to see if there are any additional changes. Record your observations in the "reaction matrix".
    With a fresh set of solutions 1-5 in clean test tubes (or wells), test each with reagent B. Repeat with reagent C.

3. *Identification of unknown*

    Obtain an unknown for Part B from your instructor. Perform the three reagent tests on your unknown. Identify the unknown as one of the five solutions from Part B. 1 above. Record your observations.

FURTHER READING

1. http://fscimage.fishersci.com/msds/21105.htm (NaCl, last accessed July, 2019)
2. https://fscimage.fishersci.com/msds/21080.htm (Na_2CO_3, last accessed July, 2019)
3. http://www.labchem.com/tools/msds/msds/LC16490.pdf ($MgSO_4$, last accessed July, 2019)
4. https://fscimage.fishersci.com/msds/01170.htm (NH_4Cl, last accessed July, 2019)
5. https://fscimage.fishersci.com/msds/20810.htm ($AgNO_3$, last accessed July, 2019)
6. http://www.labchem.com/tools/msds/msds/LC23900.pdf (NaOH, last accessed July, 2019)
7. http://www.labchem.com/tools/msds/msds/LC15320.pdf (HCl, last accessed July, 2019)
8. Selco, J., M. Bruno, and S. Chan, "Discovering Periodicity: Hands-On, Minds-On Organization of the Periodic Table by Visualizing the Unseen." *J. Chem. Educ.*, **90** (2013) 995-1002.
9. DeMeo, S., "Synthesis and Decomposition of Zinc Iodide: Model Reactions for Investigating Chemical Change in the Introductory Laboratory." *J. Chem. Educ.*, **72** (1995) 836.

Name_____ Lab Section _____ Date_____   49

Prelaboratory Assignment: Experiment 2
Identification of Compounds through Physical and Chemical Changes
Where appropriate, answers should be given to the correct number of significant digits.

1. Identify at least five different observations that are indicative of a chemical reaction.

2. Referring to a solubility table, identify the substances that are soluble and those that are not soluble in water.

   $AgNO_3$ _____ Ag_2CO_3 _____

   NaCl _____ $MgSO_4$ _____

   AgCl _____ Ag_2SO_4 _____

   Na_2CO_3 _____ NH_4Cl _____

Experiment 2 Physical and Chemical Changes

3. Indicate the ions present in solution for each compound.

   NaCl _____

   Na_2CO_3 _____

   $MgSO_4$ _____

   NH_4Cl _____

4. Three colorless solutions in test tubes, with no labels, are in a test tube rack. Three labels lie beside the test tubes: potassium iodide, KI; silver nitrate, $AgNO_3$; and sodium sulfide, Na_2S. Place the labels on the test tubes using the three solutions present. Below are your tests:

   - A portion of test tube 1 added to a portion of test tube 3 produces a yellow silver iodide precipitate.
   - A portion of test tube 1 added to a portion of test tube 2 produces a black silver sulfide precipitate.

   Your conclusions are:

   Test tube 1 contains: _____

   Test tube 2 contains: _____

   Test tube 3 contains: _____

Experiment 2 Physical and Chemical Changes

Name_____ Lab Section_____ Date_____   51

REPORT ON EXPERIMENT 2
Identification of Compounds through Physical and Chemical Changes

Part A Reaction matrix

Test	NaCl (*aq*)	Na$_2$CO$_3$ (*aq*)	MgSO$_4$ (*aq*)	NH$_4$Cl (*aq*)	H$_2$O (*l*)	Unknown
AgNO$_3$ (*aq*)						
NaOH (*aq*)						
HCl (*aq*)						

Write formulas for the precipitates that formed in the reaction matrix above.

Test	NaCl (*aq*)	Na$_2$CO$_3$ (*aq*)	MgSO$_4$ (*aq*)	NH$_4$Cl (*aq*)	H$_2$O (*l*)	Unknown
AgNO$_3$ (*aq*)						
NaOH (*aq*)						
HCl (*aq*)						

Sample number of unknown for Part A: _____

Compound in the unknown solution: _____

Part B Reaction matrix

Solution	1	2	3	4	5	Unknown
Reagent A						
Reagent B						
Reagent C						

Sample number of unknown for Part B: _____

Compound in the unknown is the same as solution number: _____

Experiment 2 Physical and Chemical Changes

QUESTIONS *(Submit your answers on a separate sheet as necessary)*
1. Fill in the blanks

_____ properties can be observed without chemically changing matter. _____ properties describe how a substance interacts with other substances. _____ have definite shapes and definite volumes. _____ have indefinite shapes and definite volumes. _____ have indefinite shapes and indefinite volumes.

Phase changes are _____ changes. As a rule of thumb, if there is ___, ___, ___, ___, or ___, it is a good indication that a chemical change has occurred.

2. Which equation below do you think represents a physical change?
    a. $H_2O(s) + Heat \rightarrow H_2O(l)$

    b. $2 H_2(g) + O_2(g) \rightarrow 2 H_2O(g) + Heat$

    c. $H_2(g) + I_2(g) + Heat \rightarrow 2 HI(g)$

    d. $N_2(g) + 2 O_2(g) + Heat \rightarrow 2 NO_2(g)$

3. Classify each of the following as a physical change or a chemical change.
    a. Melting of ice   _____
    b. Boiling of water   _____
    c. Subliming of ice   _____
    d. Decomposing of water   _____

Experiment 2 Physical and Chemical Changes

4. In an investigation, a dripless wax candle is weighed and then lighted. As the candle burns, a small amount of liquid wax forms near the flame. After 10 minutes, the candle's flame is extinguished, and the candle is allowed to cool. The cooled candle is weighed.
   a. State *one* observation that indicates a chemical change has occurred in this investigation.

   b. Identify *one* physical change that takes place in this investigation.

5. Given the particle diagram below representing four molecules of a substance:

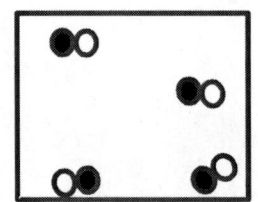

   Which particle diagram best represents this same substance after a physical change has taken place?

   a.

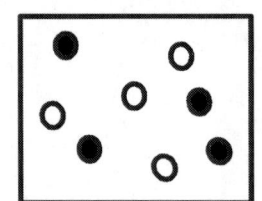

   b.

   c.

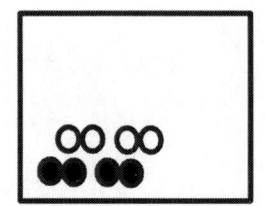

   d.

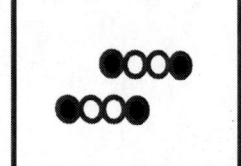

Experiment 2 Physical and Chemical Changes

6. Use the solubility chart in the Appendix to list at least 3 soluble (no exceptions) polyatomic ions.

Experiment 2 Physical and Chemical Changes

Experiment 3
Chromatography in Identification of Color Ions and Compounds

OBJECTIVES

To separate ions using paper chromatography; learn the techniques of paper chromatography: a. preparing chromatography paper to analyze samples b. setting up a developing chamber c. creating a mobile phase for a chromatography experiment; to determine the ion content of a mixture.

EQUIPMENT

Eight 12x75-mm test tubes, labels, two pieces of Whatman No. 1 filter paper cut into 10x11-cm blocks, paper towels, #2 pencil, centimeter ruler, parafilm or Saran wrap, rubber bands, two 600-mL beakers, one 150-mL beaker.

REAGENTS

$0.1M$ solutions of Fe^{3+}, Co^{2+}, Ni^{2+} and Cu^{2+} in $1M$ HCl; a mixture of Fe^{3+}, Co^{2+}, Ni^{2+} and Cu^{2+} in $1M$ HCl; several unknowns containing various combinations of these ions at $0.1M$ in $1M$ HCl; acetone; $6M$ HCl; concentrated NH_3; lecture bottle of H_2S gas or FeS_2 and concentrated H_2SO_4 for generating H_2S gas; saturated aqueous solution of Na_2CO_3.

SAFETY AND DISPOSAL

Refer to the MSDS information available online when working with $FeCl_3$,[1] $Co(NO_3)_2$,[2] $Ni(NO_3)_2$[3] or $Cu(NO_3)_2$.[4] Disposal for these compounds and their derived salts as well as Na_2CO_3, should be into a labeled laboratory waste container for inorganic chemicals and in accordance with local, state and federal regulations. Disposal for unreacted acids should be into a labeled laboratory waste jar for acids. Disposal for unreacted bases should be into a labeled laboratory waste jar for bases. Disposal for unused acetone and hydrocarbon-based solvents should be appropriately labeled waste jars for organic solvents.

INTRODUCTION

Chromatography is a physical method for separating chemical compounds. As you might imagine, the prefix chroma- indicates that color is involved. The word originated when Tswett coined it in 1903.[5] As a botanist, Tswett was interested in separating the components of various plants. Interestingly, the components had different colors and were easily identifiable and separable when passed over calcium carbonate. Color is not the only way to separate or detect the components of a mixture. There are several different chromatographic methods that are available to chemists. Gas chromatography, liquid chromatography and thin-layer chromatography are examples of some methods that are widely used. Each method has two things in common. Each one utilizes a "stationary" phase and a "mobile" phase. Usually, the phases are totally distinct from one another. For example, liquid chromatography uses a solvent as the "mobile" or moving phase and a column packed with material that has an

Fig. 3-1. TLC chamber set-up.

affinity for one or more components of the mixture that is being analyzed. The column itself is the stationary phase. For gas chromatography, the mobile phase is an inert gas like nitrogen whereas the stationary phase is a column packed with material that has an affinity for one or more components of the mixture being analyzed. Separation occurs because of several different factors. Some factors that affect how compounds separate are: size, polarity and solubility.

Thin-layer chromatography is the type of chromatography most closely related to our experiment today. An example of a thin-layer chromatography set up is shown in Fig. 3-1. Here, the mobile phase is composed of solvent vapors that come from evaporating solvent. These vapors create a capillary effect, pulling the compounds that are spotted on the TLC plate up with them. The TLC plate is coated with silica gel, an adsorbent material, and this causes compounds to "stick" or adhere to the TLC plate based on their polarity. The more polar the compound is, the shorter the distance it moves. More polar compounds stick (*i.e.* have a higher affinity for) to the plate better and move slower.

Paper chromatography uses chromatography paper as the stationary phase. The composition of the paper is unique in that it contains cellulose, a polymer. A polymer is a large molecule made up of several monomer repeat units that make up the cellulose polymer are glucose molecules. Plainly put, cellulose is made up of several glucose molecules linked together. The OH groups in cellulose are very good at "grabbing" compounds. Compounds that are attracted to the cellulose in the paper will "stick" to the paper whereas compounds that are less attracted will move up the paper much faster. A snippet of the structure of the cellulose polymer is shown below in Fig. 3-2. Remember, the OH groups in cellulose act as hands and grab compounds that they like. In more technical terms, the OH groups are excellent at hydrogen bonding. If you haven't studied hydrogen bonding yet, it is the strongest of all intermolecular forces. Some compounds bind better to OH than others and therefore will move up the chromatography paper at a much slower rate. Other compounds are more soluble in the mobile phase and move up the paper much faster.

Fig. 3-2. Cellulose structure.

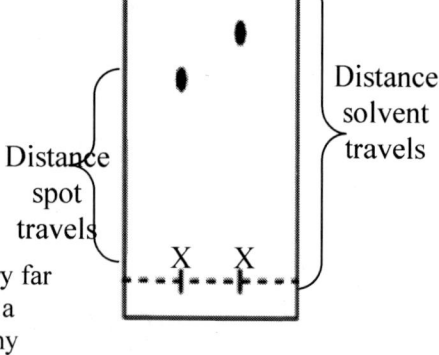

Fig. 3-3. R_f measurement.

The distance that a compound travels can be compared to the rate that it travels. If a compound travels slowly, that means that if won't move very far up the plate. If a compound travels rapidly, that means that it will move a greater distance up the plate. We compare compounds in chromatography by comparing their retardation factor (R_f) values (Fig. 3-3). R_f values are a good way to measure a compound's polarity. Highly polar compounds stick to the stationary phase and have low R_f values and the reverse is also true. R_f is readily calculated from the following equation:

$$R_f = \frac{\text{distance spot travels}}{\text{distance solvent travels}} \tag{3-1}$$

The **units for measurement are inconsequential**- cm, mm, in, may be used since they cancel out.

This experiment calls for us to analyze metal complexes via paper chromatography. Separating metal ions is a little different from separating other types of molecules. Metal complexes are either soluble in the mobile phase or they "stick" to the paper used for chromatography. If they are soluble in the mobile phase (recall, the mobile phase is responsible for "moving" the compound up the paper) then

Experiment 3 Chromatography in Identification of Color Ions and Compounds

they will move rapidly up the paper. If the complex is more attracted to the stationary phase, then it will move more slowly up the plate and have a much lower R_f. Your mobile phase for today's experiment will consist of water, acetone and HCl. Your analytes will form one of three types of complexes when exposed to this mobile phase. Aquo complexes, $M(H_2O)_x^{n+}$, will form if the analyte's counterions exchange with water. Uncharged chloro- complexes, $MCl_n(H_2O)_y$, which are also hydrates can also be formed. Negatively charged chloro- compounds are also possible and have the following molecular formula, $[MCl_{(n+y)}]^{-y}$. The application of the concept discussed earlier is as follows: If a compound forms a chloro- (neutral hydrate or negatively charged) complex it will be highly soluble in the mobile phase and, as expected, will move up the paper with ease. The aquo complexes, because of their ability to hydrogen bond, will stick more closely to the cellulose based chromatography paper (Fig. 3-4).

Fig. 3-4. Cellulose/Metal-aquo complex.

In this experiment we have an opportunity to determine the identity of unknowns based on R_f values, colors and by knowing some simple exchange reactions that take place. If we co-spot (*i.e.* spot several species on the same spot) a mixture of several different metal complexes, the components of which are known, and compare it to individual unknowns, we should be able to identify the unknowns based on what we know about the mixture. The TLC plate below is an example (Fig. 3-5).

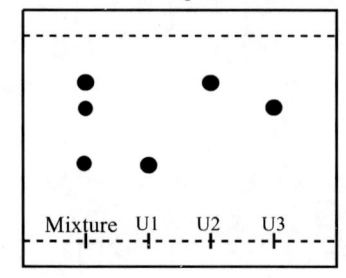

Fig. 3-5. Determining unknowns from a mixture.

PROCEDURE
Wear your safety goggles at all times in the laboratory. Remember, exercise caution when conducting experiments. Never do any experiment without the consent of the instructor.

Part A: Preparation of the solvent (Mobile Phase)
The mobile phase is very important and it changes with each type of chromatography experiment. For this experiment, the mobile phase consists of acetone and aqueous HCl.
    a. Obtain 5 mL of 6*M* HCl
    b. Obtain 18 mL of acetone
    c. Combine and mix well acetone and HCl in a 50-mL beaker.

Part B: Preparation of the Developing Chamber
To complete the chromatography experiment, the chromatography paper must be "developed". The "developing chamber" is simply a beaker with a watch glass to cover it.
    a. Obtain a 600-mL beaker and a watch glass.
    b. Add 5-10 mL of the solvent (mobile phase) prepared earlier to the beaker and cover it with the watch glass.

Part C: Preparation of chromatography paper (Stationary Phase) (Fig. 3-6)
    a. Obtain two pieces of filter paper about 11 cm x 10 cm: one for D Part 1 and one for D Part 2.
    NOTE: You can adjust the size of your filter paper to accommodate various beaker sizes.

Experiment 3 Chromatography in Identification of Color Ions and Compounds

b. Fold the paper into four equal size panels.
c. Unfold and number each panel 1-4.
d. Draw a line about **2 cm from the bottom of the paper-** and place a **small pencil mark (dot) above it.**
e. Put an X-mark in pencil in the center of each panel and slightly above the bottom line. **DO NOT USE PEN: Ink contains chemicals that will also be absorbed into your paper.**

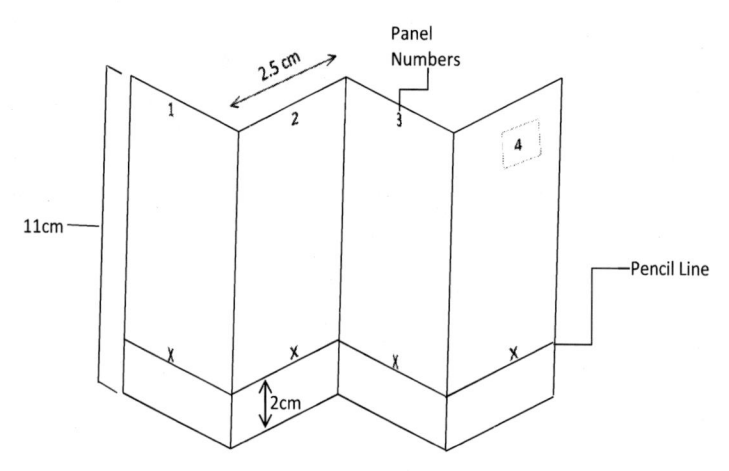

Fig. 3-6. Chromatography paper folded into panels.

Part D: Preparing the paper for development [Note: Your instructor may make stock solutions to preclude the use of multiple test-tubes.]

D. Part I
   a. Obtain 5 glass capillaries for "spotting" the solutions. Each solution should be spotted with a clean capillary.
   b. Obtain 5 test tubes: To tube 1, add the known mixture of ions (Fe^{3+}, Co^{2+}, Ni^{2+}, Cu^{2+}). To tube 2, add 1 mL of $0.1M$ Ni^{2+}. To tube 3, add 1 mL of $0.1M$ Co^{2+}. To tube 4, add 1 mL of $0.1M$ Fe^{3+}. To tube 5, add 1 mL of $0.1M$ Cu^{2+}. Also, obtain three unknown samples. Be sure to properly label your test tubes using a sharpie or a label.
   c. Dip the capillary into test tube 4 and spot Fe^{3+} on panel 1. Be sure to spot your compound just above the line (see Fig. 3-3). It is important to never let your spot touch the solvent- always spot above the 2-cm line that you drew earlier. **DO NOT SPOT compounds on top of one another or directly on the line!**
   d. Repeat Part C for Co^{2+}, Ni^{2+} and Cu^{2+}
   e. Place your chromatography paper inside the developing chamber (beaker) and cover it with a watch glass
   f. Observe the chromatography paper for any colored spots that are eluting up the paper
   g. Remove the paper when the solvent reaches 2 cm from the top of the plate
   h. Mark the line where the solvent stops with a **PENCIL. DO NOT USE PEN - THE INK WILL DISSOLVE AND SPOT ALONG WITH YOUR COMPOUND.**
   i. Circle any visible spots with your pencil
   j. Air-dry the chromatography paper by lightly waving it back and forth. Gently warming the paper with a heat gun may also be used to dry it. It should be sufficiently dry before the next step.
   k. Measure the R_f by measuring the distance from the middle of the spot to the point where the spot migrated from (Fig. 3-5). Then measure the distance from the 2 cm line to the line where the solvent stopped. R_f is a unitless measurement because the units for distance will cancel.
   l. Calculate R_f for every spot on the chromatogram.

D. Part II
   a. Obtain 3 *additional* test tubes and number them 2,3,4 to be compared with the mixture in tube 1 from Part I: Keep the test tube that contains all the metal ions (test-tube I Part I). To tube 2, add 1 mL of unknown 1. To tube 3, add 1 mL unknown 2. To tube 4, add 1 mL of unknown 3.

b. Dip the capillary into test tube 1 and spot the unknown on spot 1. Repeat for test tube 2,3,4. Be sure to spot your compound just above the line (see Fig. 3-6). **DO NOT SPOT compounds on top of one another or directly on the line!**
c. Place your chromatography paper inside the developing chamber (beaker) and cover it with a watch glass
d. Observe the chromatography paper for any colored spots that are eluting up the paper.
e. Remove the paper when the solvent reaches 2 cm from the top.
f. Mark the line where the solvent stops with a PENCIL.
g. Circle any visible spots with your pencil.
h. Calculate the R_f of each spot.
i. Air-dry the chromatography paper by lightly waving it back and forth. Gently warming the paper with a heat gun may also be used to dry it. It should be sufficiently dry before the next step.

Part E: Visualizing Spots with Color

Many times, spots on the solid phase of a paper chromatography experiment are invisible and must be "visualized" using a reagent that has an affinity for the compound(s) that are present on the paper. For instance, ninhydrin is a visualization agent that is specific for nitrogen containing compounds such as amines. When the stationary phase where amines are present is exposed to ninhydrin, it turns the amines blue. Ammonia can also be used as a visualization reagent. It is able to react with metal-aquo complexes by doing ligand exchange (i.e. substituting for a water molecule) or by deprotonation (Fig. 3-7).

$$[M(H_2O)_6]^{2+} + NH_3 \rightleftarrows [M(H_2O)_5(OH)]^{+} + NH_4^{+} \textbf{ Deprotonation}$$

$$[M(H_2O)_6]^{2+} + 3NH_3 \rightleftarrows [M(H_2O)_3(NH_3)_3]^{2+} + 3H_2O \textbf{ Ligand exchange}$$

Fig. 3-7. Reaction of ammonia with charged metal-aquo complexes.

I. **Procedure for Visualization using Ammonia**
   a. Dry chromatography papers.
   b. Place a small beaker with concentrated ammonia inside a larger beaker
   c. Place your papers inside the beaker and seal the beaker using saran wrap or parafilm and a rubber band. NOTE (If white NH_4Cl fumes form, you need to remove your papers and dry them more. NH_4Cl is a result of HCl from the mobile phase reacting with the ammonia reagent.)
   d. After 5-10 minutes of exposure, observe for any color changes.

II. **(Optional) Procedure for visualization using hydrogen sulfide (NOTE: This test should be done in a fume hood. Hydrogen sulfide gas is noxious and can be an irritant.)**
   a. Dry chromatography papers.
   b. Place the papers into a 1000-mL beaker and place the beaker into the fume hood
   c. From a lecture-bottle (gas cylinder) or an H_2S generator, pass H_2S gas over the papers.
   d. Seal the beaker using saran wrap or parafilm and a rubber band.
   e. After 5-10 minutes of exposure, observe for any color changes.

FURTHER READING
1. http://www.labchem.com/tools/msds/msds/LC14290.pdf ($FeCl_3$, last accessed July, 2019).
2. https://fscimage.fishersci.com/msds/05290.htm ($Co(NO_3)_2$, last accessed July, 2019).
3. https://fscimage.fishersci.com/msds/16370.htm ($Ni(NO_3)_2$, last accessed July, 2019).
4. https://fscimage.fishersci.com/msds/45469.htm ($Cu(NO_3)_2$, last accessed July, 2019).
5. Beran, J.A., Laboratory Manual for Principles of General Chemistry, Wiley and Sons, (1994) 189.
6. Nagel, M.L.; "Visualizing Separations: How Shopping Can Be Useful for Introducing Chromatography" *J. Chem. Educ.*, **90** (2013) 93.
7. Johnston, A.; Scaggs, J.; Mallory, C.; Haskett, A.; Warner, D.; Brown, E.; Hammond, K.; McCormick, M.M.; McDougal, O.M.; "A Green Approach to Separate Spinach Pigments by Column Chromatography" *J. Chem. Educ.*, **90** (2013) 796.
8. Clay, M.D.; McLeod, E.J.; "Detection of Salicylic Acid in Willow Bark: An Addition to a Classic Series of Experiments in the Introductory Organic Chemistry Laboratory" *J. Chem. Ed.* **89** (2012) 1068.
9. Brozek, C.M.; "Chromatography" *J. Chem. Educ.*, **76** (1999) 83.
10. Still W.C.; Kahn, M.; Mitra, A.; "Rapid Chromatographic Technique for Preparative Separations with Moderate Resolution" *J. Org. Chem.* **1978**, *43*, 2923.

Name _____ Lab Section _____ Date_____   61

Prelaboratory Assignment: Experiment 3
Chromatography in Identification of Color Ions and Compounds

1. Consider the TLC plate shown below. A mixture of compounds was spotted on the plate.
   a. How many components are in the mixture?

   b. Which spot is most polar?

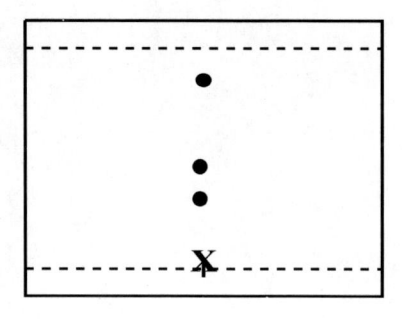

   c. Which spot is least polar?

2. What is the stationary phase for each of the following chromatographic techniques?
   a. Gas chromatography

   b. Liquid chromatography

   c. Supercritical fluid chromatography

   d. Ion exchange chromatography

   e. Preparative thin-layer chromatography

Experiment 3 Chromatography in Identification of Color Ions and Compounds

3. Define "eluent" as it relates to paper chromatography.

4. What does ammonia do to metal aquo complexes?

5. What intermolecular force is responsible for the attraction between the metal aquo complex and the cellulose form which the chromatography paper is made.

Name _____ Lab Section _____ Date _____ 63

REPORT ON EXPERIMENT 3
Chromatography in Identification of Color Ions and Compounds

Identity of ion	Spot # on Plate	Color after drying	Color after exposure to NH_3	Color after exposure to H_2S (optional)	Distance spot moved up paper	Distance Solvent moved	R_f values

Unknown #	Cations Present in Mixture

Show R_f calculations here (*attach additional sheets if necessary*)

Distance solvent traveled from 2-cm line	
Chromatogram 1: _____ cm	**Chromatogram 2:** _____ cm

Experiment 3 Chromatography in Identification of Color Ions and Compounds

QUESTIONS

(Submit your answers on a separate sheet as necessary.)

1. Which ion produced a dark blue color when exposed to NH_3? Write the chemical equation for the reaction that occurred.

2. Why is it important to keep the solvent below the 2-cm line at the bottom of the chromatography paper?

3. When detecting the identity of an unknown, it is useful to compare it to a known solution. Is it necessary to develop the chromatograms for the same amount of time? Explain why.

4. Based on the R_f values calculated, which of your ions existed mainly as an aquo complex? Explain your answer.

Experiment 3 Chromatography in Identification of Color Ions and Compounds

Experiment 4
Introduction to Gravimetric Analysis

OBJECTIVE

To determine gravimetrically the water of crystallization in a salt hydrate.

EQUIPMENT

Digital balance, two porcelain crucibles and covers, ring stand with 3- in. iron ring, wire triangle, Bunsen burner, weighing paper, spatula, crucible tongs.

REAGENTS

2.0 g of salt hydrate.

SAFETY AND DISPOSAL

Refer to the MSDS information available online when working with $CuSO_4$,[1] $CaCl_2$,[2] $NaC_2H_3O_2$,[3] $MgSO_4$,[4] $BaCl_2$,[5] $CaCO_3$,[6] and $CaSO_4$.[7] Disposal for these compounds should be into a labeled laboratory waste container for inorganic chemicals and in accordance with local, state and federal regulations.

INTRODUCTION[8]

Many crystalline materials commonly found in nature are stoichiometric combinations of two or more compounds, each of which is capable of existing independently. Often, removing one of these compounds alters the crystallinity of the materials. For example, Epsom salt is a crystalline substance with the formula $MgSO_4 \cdot 7H_2O$. Each magnesium sulfate formula unit is combined with seven molecules of water. When Epsom salt is gently heated at about 200°C, the water is driven away, leaving magnesium sulfate as an *anhydrous* (without water) powder.

The quantity of any compound having a mass in grams equal to the formula weight of the compound contains 1 mol of formula units. Thus, 1 mol of Epsom salt has a mass of 246 g. In one formula unit of $MgSO_4 \cdot 7H_2O$ there is 1 mol of $MgSO_4$ with a mass of 120 g and 7 mol of H_2O with a mass of 126 g. A mass of 246 g of Epsom salt should reduce to 120 g of anhydrous powder when heated gently.

Crystalline compounds such as Epsom salt, which have definite amounts of water associated with each formula unit of inorganic material in the crystal, are classified as *hydrates*.

The general formula for most salt hydrates is $A_mB_n \cdot xH_2O$, where m, n, and x are integers. In this experiment your instructor will tell you the identity of the salt, A_mB_n, which is assigned to you. You will then heat a weighed sample of the hydrated salt to drive off the water of crystallization and weigh the remaining anhydrous salt. These data will enable you to find the number of moles of water, x, per mole of anhydrous salt, A_mB_n. This method is an example of *gravimetric analysis* (analysis through the measurement of mass).

PROCEDURE

Wear your safety goggles throughout the experiment. Review the material on the analytic balance and use of the Bunsen burner in the *Laboratory Equipment and Techniques* section before attempting this procedure. All calculations must be performed to the proper number of significant digits.

Carefully clean a crucible and crucible cover and wipe them with a dry towel. Set up a ring stand as shown in the Fig. 4-1. Place the crucible, with the cover slightly ajar, on the supporting triangle. Heat the crucible to dull red in the hottest portion of the flame of your laboratory burner.

Ignition of the crucible is a standard analytical technique used to remove any absorbed dust, moisture, or oils. Allow at least 10 minutes for the crucible to cool to room temperature. Using either crucible tongs or weighing paper, transfer the cooled crucible and cover to an analytical balance. Do not use your fingers; they may transfer oil and moisture to the crucible. With the analytical balance, determine the mass of the crucible and cover. Record the result on the data sheet. The mass of a container (the crucible and its cover, in this case) is traditionally referred to as the *tared mass*.

Use the digital balance to weigh about 0.7 g of your assigned hydrated salt on a piece of weighing paper. Transfer the crystals to the crucible, replace its cover, and reweigh on the analytical balance. Record this mass. Determine the mass of hydrated salt crystals by subtracting the tared mass. Record the results on the data sheet. Place the crucible as before, with its cover slightly ajar, in the wire triangle. Gently heat the crucible for 15 minutes with a low flame containing a single blue zone. To avoid decomposition of the anhydrous salt, *do not overheat.* Put the crucible cover into a fully seated position, using the crucible tongs, and then turn off the flame.

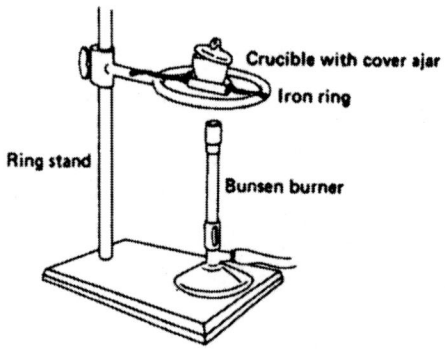

Fig. 4-1. Apparatus set-up

While the covered crucible and residue are cooling, clean and dry a second crucible and cover to prepare for a second trial determination. Carefully reweigh the covered crucible and contents after they have cooled to room temperature. Record the mass on your report sheet. Determine the mass of the anhydrous salt by subtracting the tared mass from the mass just recorded.

Ignite the second crucible and cover. Repeat the procedure with a fresh sample of salt hydrate. To clean the crucibles, add drops of water until the anhydrous powder dissolves. Record any observations, such as evolution of heat or a color change, which might suggest a chemical reaction. Pour the resultant solutions into the appropriate laboratory waste jar.

Optional Work. If time permits, obtain a different hydrate salt from your instructor and analyze it in the manner described above. Set up your own data entries and submit them on a separate report sheet.

Calculations. After recording the data collected from a procedure similar to the one described here, a student obtained the following results from gravimetric analysis of the water of crystallization of Epsom salt

mass of hydrated $MgSO_4$	= 0.7034g
mass of anhydrous $MgSO_4$	= 0.3431g
mass of water lost	= 0.3603 g

Note that the sum of the mass of the anhydrous material and the water lost equals the mass of the hydrated material. The number of moles of anhydrous $MgSO_4$ is calculated using its formula weight.

$$0.3431 \text{g } MgSO_4 \times \frac{1.000 \text{ mol}}{120.4 \text{g } MgSO_4} = 2.850 \times 10^{-3} \text{ mol } MgSO_4$$

Experiment 4 Introduction to Gravimetric Analysis

Similarly, the number of moles of water lost is:

$$0.3603 \text{ g H}_2\text{O} \times \frac{1.000 \text{ mol}}{18.02 \text{ g H}_2\text{O}} = 1.999 \times 10^{-2} \text{ mol H}_2\text{O}$$

The ratio of the number of moles of H$_2$O to moles of MgSO$_4$ is

$$\frac{1.999 \times 10^{-2} \text{ mol H}_2\text{O}}{2.850 \times 10^{-3} \text{ mol MgSO}_4} = 7.014 \text{ mol H}_2\text{O/mol MgSO}_4$$

Thus, the formula for hydrated magnesium sulfate is MgSO$_4$•7H$_2$O.

FURTHER READING
1. https://fscimage.fishersci.com/msds/05690.htm (CuSO$_4$, last accessed July, 2019)
2. https://fscimage.fishersci.com/msds/03900.htm (CaCl$_2$, last accessed July, 2019)
3. https://fscimage.fishersci.com/msds/20860.htm (NaC$_2$H$_3$O$_2$, last accessed July, 2019)
4. http://www.labchem.com/tools/msds/msds/LC16490.pdf (MgSO$_4$, last accessed July, 2019)
5. https://fscimage.fishersci.com/msds/02371.htm (BaCl$_2$, last accessed July, 2019)
6. https://fscimage.fishersci.com/msds/03880.htm (CaCO$_3$, last accessed July, 2019)
7. https://esciencelabs.com/sites/default/files/msds_files/Calcium%20Sulfate.pdf (CaSO$_4$, last accessed July, 2019)
8. Hall, C. "On the History of Portland Cement After 150 Years." *J. Chem. Educ.* **53** (1976) 222.

COMMENTS

Name _____ Lab Section _____ Date _____   69

Prelaboratory Assignment: Experiment 4
Gravimetric Analysis Water of Hydration and Crystallization
Where appropriate, answers should be given to the correct number of significant digits.

List the following in order by the greatest percentage loss of mass upon heating to an anhydrous compound? Show your calculations to receive full credit.

a. $CuSO_4 \cdot 7H_2O$

b. $CaCl_2 \cdot H_2O$

c. $NaC_2H_3O_2 \cdot 5H_2O$

d. $MgSO_4 \cdot 6H_2O$

e. $BaCl_2 \cdot 3H_2O$

f. $CaCO_3 \cdot 8H_2O$

g. $CaSO_4 \cdot 4H_2O$

Experiment 4 Introduction to Gravimetric Analysis

If you fail to pre-heat your crucible, and the mass of crucible is 55.0g but contains 10.0g of water, what is the true mass of the crucible?

How much excess water is in the crucible?

Will this affect your water of hydration calculation?

If you fail to heat your salt to the point that it is anhydrous, what effect will this have on calculating the waters of hydration? Explain.

What physical (e.g. heat, color change, transition of solid into solution) changes do you expect to observe when you add water to an anhydrous compound?

Name _____ Lab Section _____ Date _____   71

REPORT ON EXPERIMENT 4
Introduction to Gravimetric Analysis

Assigned Salt _____

Formula Weight _____

DATA AND RESULTS

mass of crucible, cover, and hydrated salt	_____	_____
tared mass	_____	_____
mass of hydrated salt	_____	_____
mass of crucible, cover, and residue after heating	_____	_____
mass of anhydrous salt	_____	_____
moles of anhydrous salt	_____	_____
mass of water lost	_____	_____
percent by mass of water in hydrate	_____	_____
moles of water lost	_____	_____
mole ratio of H₂O to anhydrous salt	_____	_____
probable formula for hydrate	_____	_____

Show the details of all calculations: use extra sheets as necessary.

QUESTIONS
(Submit your answers on a separate sheet as necessary.)
Where appropriate, answers should be given to the correct number of significant digits.

1. A sample of $Na_3PO_4 \cdot xH_2O$ was analyzed to contain 70% Na_3PO_4 by mass.
   a. What is the percent water in the sample?

   b. What is the number of moles of the anhydrous salt?

   c. Determine the number of water molecules found in the original salt?

Experiment 4 Introduction to Gravimetric Analysis

2. Explain what physical changes are observed if a colored hydrated salt ($CuSO_4 \cdot 5H_2O$) is heated to temperatures above 200 degrees? What happens to its anhydrous salt when exposed to water vapor overnight?

3. What would you observe in the salt if the heating was done with a red-yellow flame instead of a blue flame? Provide a chemical equation to support your answer.

Experiment 4 Introduction to Gravimetric Analysis

Experiment 5
Acids, Bases, Salts, and pH

OBJECTIVE
To become familiar with the chemical properties of acids, bases, and salts. To develop the concept of pH and approximate the pH of common acids, bases, and salts.

EQUIPMENT
Part A: Six 12 x 75-mm test tubes, test tube rack, test tube holder, marker (to label test tubes), pH paper.
Part B: dropper bottle.
Part C: Six 12 x 75-mm test tubes, pH paper.
Part D1: Three 12 x 75-mm test tubes. *Part D2*: Nine 12 x 75-mm test tubes. *Part D3*: Six x 75-mm test tubes.

REAGENTS
Part A: Ten drops each of 0.2 M aqueous solutions of HCl, HNO$_3$, NH$_4$Cl, NH$_3$, and NaOH.
Part B: 3 mL 0.2 M NaOH.
Part C: Ten drops each 0.2 M aqueous solutions of FeCl$_3$, NH$_4$Cl, NaCl, Na$_2$CO$_3$, Na$_3$PO$_4$, and NaH$_2$PO$_4$.
Part D1: One 0.5 x 1" strip each Zn, Mg, and Cu; 6M HCl. *Part D2*: Nine 0.5 x 1" strips Zn metal; 3M, 0.5M, and 0.01M of H$_3$PO$_4$, HCl, and H$_2$SO$_4$. *Part D3*: Three 0.5 x 1" strips each Mg and Cu, 3M H$_3$PO$_4$, 3M H$_2$SO$_4$, and 3M HCl. Red sorghum indicator (or phenolphthalein indicator).

Natural indicator preparation[1]
Weigh approximately 1.00 g of powdered sample leaves (≤ 2 mm mesh sized) into a Pyrex culture test tube (25 × 200-mm) and add 25.0 mL of ethanol (99.9%).[2] The mixture should be vortexed for 5 minutes at ambient temperature (25°C) and then filtered using Whatman No. 4 filter paper into a new culture test tube, capped and ready for use on the same day. **Caution:** The powder should be sieved (≤ 2mm mesh) into an amber bottle and stored away from direct sun-light to prevent photolysis and decomposition prior to preparation of the indicator.

SAFETY AND DISPOSAL
Refer to the MSDS information available online when working with HCl,[3] HNO$_3$,[4] NH$_4$Cl,[5] NH$_3$,[6] NaOH,[7] FeCl$_3$,[8] NaCl,[9] Na$_2$CO$_3$,[10] Na$_3$PO$_4$,[11] NaH$_2$PO$_4$,[12] H$_3$PO$_4$,[13] and H$_2$SO$_4$.[14]

Disposal for all compounds must be in accordance with local, state and federal regulations. Disposal for NH$_4$Cl, FeCl$_3$, NaCl, Na$_2$CO$_3$, Na$_3$PO$_4$, NaH$_2$PO$_4$, should be into a labeled laboratory waste container for inorganic chemicals. Disposal for HCl, HNO$_3$, H$_3$PO$_4$, and H$_2$SO$_4$ should be into a labeled laboratory waste container for acids. Disposal for NH$_3$ and NaOH should be into a labeled laboratory waste container for bases.

INTRODUCTION
Almost every liquid you encounter has either acidic or basic traits. What does it mean for a solution to be *acidic* or *basic*? It all has to do with hydrogen ions (H$^+$). Let's start our discussion with water, H$_2$O. Water is amazing because it is considered both an acid and a base depending on how you

look at it! In H_2O, a small number of the molecules dissociate (split up). Some of the water molecules lose a hydrogen ion and become hydroxide ions (OH^-). Those "lost" hydrogen ions join up with other water molecules to form hydronium ions (H_3O^+) as shown in the chemical reaction below:

$$2\ H_2O \rightleftarrows H_3O^+ + OH^- \tag{5-1}$$

(Note: For simplicity, H_3O^+ ions are usually referred to as H^+, also called a proton)

This reaction is called the *autoionization* of H_2O. In pure water, there are an equal number of H^+ and OH^- ions, making the solution neither acidic nor basic. When the balance of these ions is *not* equal they will shift the solution to either become more acidic or more basic. In other words, when an H^+ ion is released into a system, the solution becomes acidic. When an OH^- ion is released, the solution becomes basic. Those two special ions determine whether you are looking at an acid or a base.

Acids/Bases: Often acids are considered substances that ionize in H_2O to produce H^+ ions. A Swedish chemist named Svante August Arrhenius in 1887 created this definition of acids. This definition is limited to aqueous solutions (those only dissolved in H_2O). So, a Danish chemist, Brønsted, came up with a broader definition! He proposed that an ***acid is a proton donor***. Hydrochloric acid is a Brønsted acid because it donates a proton to H_2O:

$$HCl(aq) + H_2O(l) \rightarrow H_3O^+(aq) + Cl^-(aq) \tag{5-2}$$

On the other hand, bases are substances that ionize in H_2O to produce OH^- ions. According to Brønsted, a ***base is a proton acceptor***. Ammonia is a Brønsted base because it can accept an H^+ ion:

$$NH_3(aq) + H_2O(l) \rightleftarrows NH_4^+(aq) + OH^-(aq) \tag{5-3}$$

There is a table at the end of the introduction that summarizes the properties of Brønsted acids and bases.

Acid-Base Neutralization: What happens when acids react with bases? Generally, aqueous acid-base reactions produce water and a **salt** *(ionic compound made up of a cation other that H^+ and an anion other than OH^-).*

$$Acid + Base \rightarrow Salt + Water$$

The substance we know as table salt is the product of an acid-base neutralization:

$$HCl(aq) + NaOH(aq) \rightarrow NaCl(aq) + H_2O(l) \tag{5-4}$$

Salts can be formed in many other ways. One way that you will encounter in this experiment is when you dissolve a metal in an acid. When this happens, a salt is formed along with hydrogen gas (H_2). In the example below magnesium metal dissolves in sulfuric acid to give magnesium sulfate and hydrogen gas

$$H_2SO_4(aq) + Mg(s) \rightarrow MgSO_4(aq) + H_2(g) \tag{5-5}$$

Conjugate Acids and Bases: In an acid-base reaction, an acid plus a base reacts to form a conjugate base plus a conjugate acid.

$$Acid + Base \rightarrow Conjugate\ Base + Conjugate\ Acid$$

In eqn. (5-3) above where ammonia (NH_3) accepts a proton from water we say that NH_4^+ is the **conjugate acid** to the base NH_3, because NH_3 gained a hydrogen ion to form NH_4^+, the conjugate acid. The

conjugate base of an acid is formed when the acid donates a proton. In the same equation, OH⁻ is the conjugate base to the acid H₂O, because H₂O donates a hydrogen ion to form OH⁻, the conjugate base. **Note**: *the stronger the acid or base, the weaker the conjugate; the weaker the acid or base, the stronger the conjugate.*

pH: The **pH** scale measures the strength of an acidic or basic liquid. Although there may be many types of ions in a solution, pH only focuses on concentrations of H^+ and OH^-. The scale measures values from 0 to 14. Distilled water is 7 (right in the middle; neutral). Acids are found between 0 and 7 (where concentration of H^+ ions is more abundant). Bases are from 7 to 14 (where concentration of OH^- ions is more abundant). There are also very strong acids with pH values below 1, such as battery acid. Bases with pH values near 14 include drain cleaner and sodium hydroxide (NaOH). One way to measure pH is to use an acid-base indicator. Litmus is an example of this; it is a dye molecule that changes to blue if *very few H^+* ions are present and turns red if *too many H^+* ions are present (or very few OH⁻). Another example is Universal indicator. Universal indicator contains several different acid-base indicator compounds. Each component changes color at different concentrations of acids or bases. Thus, it can measure the strength of an acid or base. The following shows the relationship between the Universal Indicator color and the degree of acid or base strength:

most acidic → somewhat acidic → somewhat basic → most basic
red → orange → yellow → green → blue → purple

Acids	Bases
Donate protons	Accept protons
Sour taste	Bitter taste
Neutralize bases	Neutralize acids
Strength depends on H^+ concentration	Strength depends on OH^- concentration
Turn blue litmus red	Turn red litmus blue
React with metals to release H_2	

Table 5-1. Properties of Brønsted Acids and Bases

PROCEDURE
Wear your safety goggles at all times in the laboratory. You will be working with highly concentrated acids. Acids and bases are potentially dangerous in high concentrations and you should exercise caution when conducting this experiment. Remain alert at all times.

Part A: pH testing of solutions to predict acidity or basicity.
Estimate the pH of each one of your solutions on the data sheet. Indicate which ones you think are acids and which ones you think are bases. Obtain 6 clean test tubes from your instructor; label them as HCl, HNO₃, H₃PO₄, NH₄Cl, NH₃ and NaOH.

Obtain 10 drops of 0.2*M* HCl solution in the test tube labeled HCl. Use the other labeled test tubes and add the 10 drops of the specified solution to them as well. DO NOT MIX THE SOLUTIONS. Obtain pH paper from your lab instructor. pH paper allows for you to estimate the acidity or basicity of a given solution based on color. Acids range in pH from 0-6, bases range in pH from 8-14. Dip the pH paper into each one of the solutions and record your results on the data sheet provided.

Part B: Neutralizing Acids.
Predict how many drops of 0.2*M* NaOH needed to neutralize **each of the 10 drops of acid from part A**. On the data sheet predict whether stronger acids will require more or less NaOH to be neutralized. Add a small amount of red sorghum indicator to each of the acids from part A. Red sorghum

Experiment 5 Acids, Bases, Salts and pH

is a natural indicator that can be used to determine the end point of an acid-base reaction. Red sorghum indicator is not harmful to the environment. If red sorghum is unavailable, use phenolphthalein.

Obtain 3 mL 0.2M NaOH and a dropper bottle. Add NaOH drop wise and count the number of drops required for the solution to change color. Record your results on Section B of the data sheet provided. How did your predictions compare to the results?

Part C: pH of Metal Salts.

Recall: Strong Acid + Strong Base = Salt + Water. Recall also that a Brønsted acid reacts with a Brønsted base, to simultaneously form a conjugate acid (the base after it accepts H^+) and a conjugate base (the acid after it loses H^+). A salt is the result of neutralization, *e.g.*, bringing the pH to 7, in the reaction of a *strong* acid with a *strong* base. Predict on your data sheets whether salts have acidic or basic pH.

Obtain 6 clean test tubes and label them Na_2CO_3, $NaHCO_3$, $NaCl$, $FeCl_3$, Na_3PO_4, NaH_2PO_4. Obtain 10 drops of 0.2M $FeCl_3$ and add it to the test tube labeled $FeCl_3$. Use the other labeled test tubes and add the 10 drops of the specified solution to them as well. DO NOT MIX THE SOLUTIONS.

Obtain pH paper from your lab instructor. Dip the pH paper into each one of the solutions and record your results on the data sheet provided. How did your predictions compare to the results?

Part D: Reaction of Metals with Acid.

Metals react with acids by a displacement reaction, producing metal ions and hydrogen gas. You can observe this in the reaction by the amount of "bubbling" that takes place when the metal is treated with an acid. The "bubbles" are hydrogen gas escaping the solution.

The reaction is as follows:

$$M + 2HX \rightarrow MX_2 + H_2 \,(gas) \qquad (5\text{-}6)$$

D1. Reaction of different metals with the same acid

Obtain three test tubes and label them zinc (Zn), magnesium (Mg), and copper (Cu). On your data sheet, predict which metal will react with HCl. Obtain a strip of Zn and place it into the test tube labeled Zn. Use the other labeled test tubes and add the specified metal to them.

Add enough 6M HCl to each test tube to submerge the metal strips and observe over several minutes. *CAUTION: 6M HCl is corrosive and should be handled with care.* Record any observations on your data sheet.

D2. Reacting Zn metal with different acids of different concentration

Obtain nine test tubes and label all of them Zn. Obtain nine strips of Zn and add one strip to each of the test tubes labeled Zn. Obtain 3M H_3PO_4, 0.5M H_3PO_4 and 0.01M H_3PO_4. Mark three of the test tubes 3M, 0.5M and 0.01M.

To the Zn test tube marked 3M, add enough 3M H_3PO_4 to submerge the metal strip and observe. Repeat this process for 0.5M and 0.01M H_3PO_4.

Use three more test tubes and repeat the above process with 3M, 0.5M and 0.01M HCl. Use three more test tubes and repeat the above process with 3M, 0.5M and 0.01M H_2SO_4.

Observe and compare each Zn test tube with each acid at the three different acid concentrations. Record any observations on your data sheet. Which acid concentration did you think would cause the metal to react faster? Does the reaction slow down at lower concentration? How can you tell?

D3. Comparing metal reactivity with different acids

Obtain 6 test tubes. Label three of them Mg and three of them Cu. On your data sheet, predict which metal will react fastest with which acid at which concentration.

Obtain three strips of Mg and add one strip to each of the test tubes labeled Mg. Repeat the process for Cu.

Obtain $3M$ H_3PO_4, $3M$ H_2SO_4 and $3M$ HCl. Mark one of the three of the test tubes H_3PO_4, one of them H_2SO_4 and the other HCl. To the Mg test tube marked H_3PO_4, add enough $3M$ H_3PO_4 to submerge the metal strip and observe. Repeat this process for the Mg tubes marked H_2SO_4 and HCl. Which acid do you predict to have the highest rate of reaction? Record any observations on your data sheet.

Repeat the above process for Cu. Record any observations on your data sheet. Comparing all three metals, which one reacted fastest with which acid at $3M$? Did the results match up with your predictions?

FURTHER READING
1. Abugri, D. A.; Apea, O. B.; and Pritchett G.; "Investigation of a Simple and Cheap Source of a Natural Indicator for Acid-Base Titration: Effects of System Conditions on Natural Indicators" *Green and Sustainable Chemistry*, (2012), **2**, 117-122.
2. http://www.sigmaaldrich.com/MSDS/MSDS/DisplayMSDSPage.do?country=US&language=en&productNumber=E7023&brand=SIAL&PageToGoToURL=http%3A%2F%2Fwww.sigmaaldrich.com%2Fcatalog%2Fproduct%2Fsial%2Fe7023%3Flang%3Den (absolute ethanol, last accessed 11/2013)
3. http://www.labchem.com/tools/msds/msds/LC15300.pdf (HCl, last accessed July, 2019)
4. http://www.labchem.com/tools/msds/msds/LC17870.pdf (HNO_3, last accessed July, 2019)
5. https://fscimage.fishersci.com/msds/01170.htm (NH_4Cl, last accessed July, 2019)
6. https://fscimage.fishersci.com/msds/00211.htm (NH_3, last accessed July, 2019)
7. http://www.labchem.com/tools/msds/msds/LC24350.pdf (NaOH, last accessed July, 2019)
8. http://www.labchem.com/tools/msds/msds/LC14380.pdf ($FeCl_3$, last accessed July, 2019)
9. https://fscimage.fishersci.com/msds/21105.htm (NaCl, last accessed July, 2019)
10. https://fscimage.fishersci.com/msds/21080.htm (Na_2CO_3, last accessed July, 2019)
11. https://fscimage.fishersci.com/msds/24500.htm (Na_3PO_4, last accessed July, 2019)
12. http://www.labchem.com/tools/msds/msds/LC24775.pdf (NaH_2PO_4, last accessed July, 2019)
13. http://www.labchem.com/tools/msds/msds/LC18640.pdf (H_3PO_4, last accessed July, 2019)
14. https://fscimage.fishersci.com/msds/22350.htm H_2SO_4, last accessed July, 2019)
15. Michałowski, T.; Asuero, A.G.; Wybraniec, S.; "The Titration in the Kjeldahl Method of Nitrogen Determination: Base or Acid as Titrant?" *J. Chem. Educ.*, **90** *(*2013*)* 191
16. Kolb, D.; "Acids and Bases" *J. Chem. Educ.*, **55** (1978) 459
17. Laredo, T.; "Changing the First-Year Chemistry Laboratory Manual to Implement a Problem-Based Approach That Improves Student Engagement" *J. Chem. Educ.*, **90** *(*2013*)* 1151

COMMENTS

Name _____ Lab Section _____ Date _____ 79

Prelaboratory Assignment: Experiment 5
Acids, Bases, Salts, and pH

Where appropriate, answers should be given to the correct number of significant digits.

1. Using the reactions shown, classify the substance in **bold** as a Brønsted **A**cid, Brønsted **B**ase, or **S**alt.

Reaction	A, B, or S
NH_4^+ (aq) + H_2O (l) → $H_3O^+(aq)$ + $NH_3(aq)$	
HNO_3 (aq) + H_2O (l) → $H_3O^+(aq)$ + $NO_3^-(aq)$	
$HCl(aq)$ + $NaOH(aq)$ → **$NaCl$**(aq) + $H_2O(l)$	
NH_3 (aq) + $H_2O(l)$ ⇌ $NH_4^+(aq)$ + $OH^-(aq)$	
$H_2CrO_4(aq)$ + **$NaOH$**(aq) → $Na_2CrO_4(aq)$ + $H_2O(l)$	

2. For the following acids, balance the complete neutralization reactions with NaOH

   ____$HCl(aq)$ + ____$NaOH(aq)$ → ____$NaCl(aq)$ + ____$H_2O(l)$

   ____$H_2SO_4(aq)$ + ____$NaOH(aq)$ → ____$Na_2SO_4(aq)$ + ____$H_2O(l)$

   ____$NH_4Cl(aq)$ + ____$NaOH(aq)$ → ____$NaCl(aq)$ + ____$NH_3(l)$ + ____$H_2O(l)$

   ____$H_2CrO_4(aq)$ + ____$NaOH(aq)$ → ____$Na_2CrO_4(aq)$ + ____$H_2O(l)$

3. A student uses pH paper to test three unknown solutions and obtains the following results. Classify the solutions as **A**cidic, **B**asic, or **N**eutral.

Solution	Color	pH	Classification (A, B, or N)
1	Green	10	
2	Yellow	7	
3	Orange	2	

Experiment 5 Acids, Bases, Salts and pH

4. Because the HSO$_4^-$ ion is amphoteric, it can donate or accept a proton. Write the formula of its conjugate acid and the formula of its conjugate base.

5. Ten drops of 0.10 M HCl are placed in a test tube. How many drops of 0.10 M NaOH would you predict are needed to neutralize the HCl? Is your answer different for ten drops of 0.10 M H$_2$SO$_4$? Explain.

Experiment 5 Acids, Bases, Salts and pH

Name _____ Lab Section _____ Date_____   81

REPORT ON EXPERIMENT 5
Acids, Bases, Salts, and pH

Part A: pH testing

Compound	HCl(*aq*)	HNO$_3$(*aq*)	H$_3$PO$_4$(*aq*)	NH$_4$Cl(*aq*)	NH$_3$(*aq*)	NaOH(*aq*)
pH estimate						

Predict which of the above will have the highest pH:

Predict which will have the lowest:

If water was one of the choices, what would you predict for its pH:

Compound	HCl(*aq*)	HNO$_3$(*aq*)	H$_3$PO$_4$(*aq*)	NH$_4$Cl(*aq*)	NH$_3$(*aq*)	NaOH(*aq*)
pH (measured)						
Classification (A,B, or S)						

Which results varied from your predictions? _____

_____

_____

Which was the strongest acid? _____   Which was the strongest base? _____

Why would some acids have different pH's than others? _____

_____

_____

Part B: Neutralizing Acids.
Based upon your results from part A, predict how many drops of NaOH solution are needed to reach a neutral pH for your acids in the table below. Then test your predictions.

Acid						
Predicted drops						
Actual drops						

Which results varied from your predictions? _____

Experiment 5 Acids, Bases, Salts and pH

How does the pH of the acid from **Part A** compare with the amount of NaOH needed to reach a neutral pH?

_____

_____

Part C: pH of Metal Salts.

Solution	Na_2CO_3	$NaHCO_3$	$NaCl$	$FeCl_3$	Na_3PO_4	NaH_2PO_4
Prediction: A, B, or N						
Measured pH						
A, B, or N						

Do you think Na^+ and Cl^- ions affect the pH of a solution? Why or why not?

_____

_____

Part D: Reaction of Metals with Acid.

D1. Reaction of different metals with the same acid.
Do you think that Zn, Mg, and Cu will react at the same rate or at different rates with 6M HCl? _____

If you think they will react differently, which do you think will be fastest and which slowest? _____

Metal	Observations
Zn	
Mg	
Cu	

Rank the three metals in terms of the speed of reaction with HCl: _____

D2. Reacting Zn metal with different acids of different concentration.
Which acid at which concentration should react the fastest with Zn? _____

Which acid at which concentration should react the slowest with Zn? _____

Experiment 5 Acids, Bases, Salts and pH

Name _____ Lab Section _____ Date _____ 83

Record observations in the table below for the solutions' reactions with Zn metal.

Concentration	H₃PO₄	HCl	H₂SO₄
0.010 M			
0.50 M			
3.0 M			

How does changing the acid concentration affect the speed of the reaction? _____

_____

Based upon the behavior, rank the strengths of the acids. _____

D3. Comparing metal reactivity with different acids.

Predict which will be fastest (#1) and which combination will react the slowest (#6). Rank the others in between. Use results from D1 and D2 to guide your choices.

Concentration	Mg	Zn
8M H₃PO₄		
8M HCl		
8M H₂SO₄		

Record your observations when the acids and metals are combined.

Concentration	Mg	Zn
8M H₃PO₄		
8M H₂SO₄		
8M HCl		

Which results varied from your predictions? _____

_____

Experiment 5 Acids, Bases, Salts and pH

QUESTIONS

(Submit your answers on a separate sheet as necessary.)

1. A comic book villain is holding you at gunpoint and is making you drink a sample of acid. She gives you a beaker with 100 ml of a strong acid with pH=5. She also gives you a beaker of a strong base with a pH=10. You can add as much of the strong base to the strong acid as you want, and you must then drink the solution. You'd be best off trying to make the solution neutral before drinking it. How much of the base should you add? Explain your answer fully.

2. What exactly is the pH scale measuring? If a solution has a large number of these, is the pH very high or low?

3. From your lab experience, give two ways you could recognize a strong acid from a weak acid.

Experiment 5 Acids, Bases, Salts and pH

4. Would you expect Zn to react faster as a powder or in granular form? Explain why.

5. Balance the following equation, identify which is the acid and which is the base, and give the mole ratios of acid to base. $Ba(OH)_2(aq) + H_3PO_4(aq) \rightarrow$

Experiment 5 Acids, Bases, Salts and pH

6. The bicarbonate ion, HCO_3^-, can undergo different reactions depending on the conditions.
   a. Write the balanced chemical equation to show how the bicarbonate ion will react with hydrochloric acid. Label the acid, base, conjugate acid and conjugate base.

   b. Write the balanced chemical equation for the reaction of HCO_3^{-1} with water to form a basic solution containing hydroxide ions. Label the acid, base, conjugate acid and conjugate base.

   c. Write the balanced chemical reaction to show how $NaHCO_3$ will react with sodium hydroxide in a neutralization reaction.

Experiment 5 Acids, Bases, Salts and pH

Experiment 6
Stoichiometry of a Reaction

OBJECTIVES

To determine the balanced equation of a reaction by determining its stoichiometric ratio from the reaction of iron metal and copper (II) sulfate.

EQUIPMENT

Hot plate, 125- or 225-mL Erlenmeyer flask, 50- or 100-mL beaker

REAGENTS

Iron filings, 1.0 M copper (II) sulfate

SAFETY AND DISPOSAL

Refer to the MSDS information available online when working with $CuSO_4$,[1] $FeSO_4$,[2] or $Fe_2(SO_4)_3$,[3] acetone[4].

Disposal for all compounds must be in accordance with local, state and federal regulations. Disposal for $CuSO_4$, $FeSO_4$, or $Fe_2(SO_4)_3$ should be into a labeled laboratory waste container for inorganic chemicals. Disposal for acetone should be into a labeled laboratory waste container for organic chemicals.

INTRODUCTION

When metallic iron contacts a solution of copper (II) sulfate, an immediate and rapid reaction occurs, producing metallic copper. This reaction is known as a displacement reaction in that the iron displaces the copper from solution. When the reaction is completed, iron, initially metallic, becomes an ion in solution and the copper ion from the solution precipitates as a metal. A metal that displaces another metal from a solution of one of its salts is said to be "more active" than the displaced metal. Thus, iron is more active than copper. Other metals can be added to this list of "activity" by examining various displacement reactions.

The displacement of the copper (II) ion by iron is also known as an oxidation-reduction reaction. The Cu^{2+} ions pick up or gain electrons to form metallic copper atoms. In the process, the oxidation number or state of the copper is reduced from +2 in the ion to 0 in the element. Any **reduction reaction** must be accompanied by an associated **oxidation reaction** in which another atom provides the electron(s). When an atom gives up electrons, its oxidation state increases. Therefore, reduction is the decrease in oxidation number (gain of electrons) and oxidation is the increase in oxidation number (loss of electrons).

Among the most common forms of copper are the metal and the Cu^{2+} ion. In aqueous solution, iron can form two ions, Fe^{2+} {iron(II) or ferrous ion} and Fe^{3+} {iron (III) or ferric ion}. If Fe^{2+} is the product, eqn. (6-1) describes the reaction with Cu^{2+}; if Fe^{3+} is the product of the reaction then eqn. (6-2) is correct.

$$Fe\,(s) \; + \; Cu^{2+}(aq) \;\; \rightarrow \;\; Fe^{2+}(aq) \; + Cu\,(s) \tag{6-1}$$

$$2\,Fe\,(s) \; + \; 3\,Cu^{2+}(aq) \;\; \rightarrow \;\; 2\,Fe^{3+}(aq) \; + \; 3Cu\,(s) \tag{6-2}$$

In this experiment, the stoichiometry (mass relationships) of the reaction of iron and copper (II) will be studied to determine which of the two equations is correct. A known amount of iron metal will be added to a solution containing an excess of copper (II) sulfate. The copper product will be isolated, dried, and weighed. These weighings will be used to calculate the number of moles of iron reacted and the number of moles of copper formed to determine the appropriate equation. According to eqn. (6-1) the mole ratio of Fe to Cu is 1:1 while according to eqn. (6-2) the ratio is 2:3.

PROCEDURE
Wear your safety goggles at all times in the laboratory.

Weigh a *clean, dry* 100-mL (or 50-mL) beaker to the nearest milligram. Add ~0.5 g of iron filings and reweigh the beaker and contents (±0.001g). Add 15 mL of warm 1.0 *M* copper (II) sulfate solution and about 20 mL of warm distilled water to the beaker containing the iron filings. *Gentle* heating of the beaker may be necessary if the iron filings do not react completely.

When the reaction has ceased, allow the copper product to settle. If necessary, wash the walls of the beaker with a small amount of water from a wash bottle. Carefully decant (pour off the clear liquid above the solid) the solution to a waste beaker. Remove as much liquid as possible but do not allow any of the solid to be transferred. Add about 10 mL of distilled water to the copper product to wash off any excess ions that remain. Decant this wash liquid. Add another 10-mL portion of water and decant again.

Add about 10 mL of acetone to the product. Swirl the mixture and allow it to stand for a few minutes. Decant the liquid to the waste beaker. Add a second 10-mL portion of acetone, swirl and carefully decant the liquid. Acetone washes the water from the copper powder. Since acetone is extremely volatile, the residual acetone can be easily removed from the solid by gentle heating. **NOTE**: Acetone is extremely flammable and should be handled with care. Carefully warm the beaker on a hot plate or with a heat gun. Be careful to keep all acetone sources away from any flame. Warm the beaker until the copper product is thoroughly dry and moves freely.

Cool and reweigh the beaker and product (±0.001 g). Submit your sample to your instructor.

Calculations. Calculate the moles of iron reacted, the moles of copper produced and the iron to copper ratio. With this information determine the correct equation for this reaction.

Using the correct equation and the exact mass of iron used in the experiment, calculate the theoretical yield of copper metal. Calculate the percent yield in your experiment.

FURTHER READING
1. https://fscimage.fishersci.com/msds/05690.htm ($CuSO_4$, last accessed July, 2019)
2. https://fscimage.fishersci.com/msds/09870.htm ($FeSO_4$, last accessed July, 2019)
3. http://www.waterguardinc.com/files/90276683.pdf ($Fe_2(SO_4)_3$, last accessed last accessed July, 2019)
4. https://fscimage.fishersci.com/msds/00140.htm (acetone, last accessed July, 2019)
5. Jensen, W.B.; "The Origin of Stoichiometry Problems" *J. Chem. Educ.,* **2003**, *80*, 1248.
6. Umland, J.B.; "A Recipe for Teaching Stoichiometry" *J. Chem. Educ.,* **1984** ,*61*, 1036.
7. Webb, M.J.; "An Experimental Introduction to Stoichiometry" *J. Chem. Educ.,* **1981**, *58*, 192.
8. Parker, G.A.; "Stoichiometry of Redox Reactions" *J. Chem. Educ.,* **1980**, *57*, 721.

Name _____ Lab Section _____ Date _____   89

Prelaboratory Assignment: Experiment 6
Stoichiometry of Reaction

Where appropriate, answers should be given to the correct number of significant digits.

1. Write the balanced equation for the following reaction:

    $Na_2SO_4 + BaCl_2 \rightarrow NaCl + BaSO_4$

2. What is the stoichiometric ratio between $BaCl_2$ and $NaCl$?

3. What is the stoichiometric ratio between Na_2SO_4 and $BaCl_2$?

4. How many moles of $BaSO_4$ are created when 10.0 grams of Na_2SO_4 are used in the reaction?

5. How many moles of NaCl are created when 10.0 grams of Na_2SO_4 are used in the reaction?

6. If 1.0 gram of Na_2SO_4 and 5.0 grams of $BaCl_2$ are used, which reactant is limiting in the formation of NaCl?

Experiment 6 Stoichiometry

7. Which reactant is excess? Does the answer surprise you?

8. Calculate the theoretical yield in grams of NaCl if 2.0 grams of Na_2SO_4 are used?

9. What is the percent yield of NaCl based on the theoretical yield calculated above if 0.577 grams of NaCl is recovered from the reaction?

Name_____ Lab Section _____ Date_____

REPORT ON EXPERIMENT 6
Stoichiometry of a Reaction

Reaction Stoichiometry Data Sheet

Mass of beaker _____g

Mass of beaker and iron sample _____g

Volume of 1.0 M $CuSO_4$ _____mL

Mass of beaker and copper product _____g

Mass of iron used _____g

Moles of iron used _____mol

Mass of copper produced _____g

Moles of copper produced _____mol

Iron/copper mole ratio _____

Correct equation: _____

Theoretical yield of copper _____g

Actual yield of copper _____g

Percent yield _____%

Experiment 6 Stoichiometry

QUESTIONS

(Submit your answers on a separate sheet as necessary.)
Where appropriate, answers should be given to the correct number of significant digits.

1. Suppose that you use 0.75 g of iron in this experiment. What is the minimum volume of 1.5 M $CuSO_4$ solution required
    a) using eqn. (6-1)?

    b) using eqn. (6-2)?

2. If copper foil is added to a (colorless) solution of silver nitrate, the solution turns blue, while the foil becomes silvery.

    What is happening?

    Which metal is more active?

3. What would you expect to happen if copper metal is placed in iron(II) sulfate solution

4. There are a number of common experimental errors that can produce a percent yield of greater than 100%.
    Why is a yield greater than 100% unreasonable?

    List one common experimental error that may account for a high yield?

    How can you avoid this error? Improper weighing and calculation errors are not acceptable answers.

Experiment 6 Stoichiometry

Experiment 7
Limiting Reactant

OBJECTIVES

To determine the limiting reactant in a mixture of two salts; to determine the percent composition of each substance in a salt mixture.

EQUIPMENT

Two 250-mL Erlenmeyer flasks, 400-mL beaker, hot plate, pH paper, 65-mm watch glass, 3.5-in glass funnel, filter paper, glass rod, 100-mm Buchner funnel, two 12 x 75 mm test tubes.

REAGENTS

0.5M calcium chloride, 0.5M potassium oxalate, 6M ammonia, solid mixture of "unknown" composition of calcium chloride and potassium oxalate.

SAFETY AND DISPOSAL

Refer to the MSDS information available online when working with NH_3,[1] $K_2C_2O_4 \cdot H_2O$,[2] $CaC_2O_4 \cdot H_2O$,[3] $CaCl_2 \cdot 2H_2O$.[4] Disposal for all compounds must be in accordance with local, state and federal regulations. Dispose of solid $K_2C_2O_4 \cdot H_2O$, $CaC_2O_4 \cdot H_2O$, and $CaCl_2 \cdot H_2O$, including the associated filter paper, in the inorganic solid waste container. Dispose of calcium oxalate solutions in the inorganic liquid waste container. Disposal for NH_3 should be into a labeled laboratory waste container for bases.

INTRODUCTION

Every reaction that occurs between two or more chemicals is governed by the principle of limiting reactants. In theory, the extent of any reaction is controlled by the reactant that is present in the smallest amount. The limiting reactant is the barrier that controls how much product is formed in a given reaction. Once the limiting reactant is consumed, the reaction will be complete. When studying the limiting reactant it is important to understand the concept of stoichiometry *i.e.,* the mole-to-mole ratio of a given set of reactants. Let's consider a reaction A + B → products with one mole of A reacting with one mole of B to produce products. What happens when 12 moles of A are mixed with 6 moles of B? Assuming that the reaction goes to completion, it is understood that only 6 moles of A will react with the 6 moles of B that are present. Thus, 6 moles of A will remain unreacted. This example also introduces us to another concept; the reagent that is not fully consumed is said to be present in excess. Here, A is the "excess" reactant while B is the "limiting reactant" since all of it gets consumed.

In a practical sense, the limiting reactant is analogous to baking cookies where we need one dozen eggs for every one bag of flour to make 3-dozen cookies. If we have two bags of flour and only one dozen eggs, we will still only get 3-dozen cookies because the given ratio is one bag of flour per one dozen eggs. The eggs are limiting in this case while the flour is present in excess; one bag was used and one bag will be left over.

In this experiment we make a solution of the water-soluble solids calcium chloride and potassium oxalate, to determine the limiting reactant in the formation of calcium oxalate monohydrate. Calcium oxalate monohydrate is the reaction product that is insoluble in water and therefore forms a precipitate. The exact composition of the solid mixture will not be known but can and will be determined later in the experiment. The reaction is shown below (eqn. 7-1). Based on the stoichiometry of the balanced equation, calcium chloride and potassium oxalate are present in a 1:1 ratio. This means that, the reactant present in the least number of moles will be the limiting reactant. If there are more moles of calcium chloride present, then potassium oxalate will be limiting. Conversely, if there are more moles of potassium oxalate present, then calcium chloride will be limiting.

$$CaCl_2 \cdot 2H_2O(aq) + K_2C_2O_4 \cdot H_2O(aq) \rightarrow CaC_2O_4 \cdot H_2O(s) + 2KCl(aq) + 2H_2O(l) . \tag{7-1}$$

For this experiment the ionic equation is:

$$Ca^{2+}(aq) + 2Cl^-(aq) + 2K^+(aq) + C_2O_4^{2-}(aq) + 3H_2O(l) \rightarrow$$
$$CaC_2O_4 \cdot H_2O(s) + 2Cl^-(aq) + 2K^+(aq) + 2H_2O(l) . \tag{7-2}$$

The net ionic equation, in which the spectator ions have been removed, is shown in eqn. 7-3:

$$Ca^{2+}(aq) + C_2O_4^{2-}(aq) + H_2O(l) \rightarrow CaC_2O_4 \cdot H_2O(s). \tag{7-3}$$

Calcium oxalate monohydrate is stable at temperatures below the boiling point of water. However, it can be dehydrated if heated above 100°C and will form an anhydrous salt, CaC_2O_4. This is a critical piece of information considering that the mass of the anhydrous salt will be 18.0 g/mol (MM of water) less than the hydrate and will affect any calculations dependent upon this data.

What we can conclude from the balanced net ionic equation is this: for every one mole of Ca^{2+} (MM= 147.01 g/mol from $CaCl_2 \cdot 2H_2O$) that is consumed by reacting with one mole of $C_2O_4^{2-}$ (MM = 184.24 g/mol from $K_2C_2O_4 \cdot H_2O$), one mole of $CaC_2O_4 \cdot H_2O$ (MM = 146.12 g/mol) is produced. The dehydrated product, CaC_2O_4 (MM = 128.10) is only formed when the product is heated above 100°C.

The experiment will be done in two phases, the first of which will be done to calculate the mass of the $CaC_2O_4 \cdot H_2O$ precipitate that will form from the reaction of $CaCl_2 \cdot 2H_2O$ and $K_2C_2O_4 \cdot H_2O$. The resulting precipitate will be "digested" in order to ensure analytical accuracy. Digestion is the process of re-dissolving the precipitate in order to ensure that smaller particles, that may not have precipitated initially, are precipitated. The purpose is to safeguard against partial precipitation that will affect the amount of the precipitate collected and also affect the percent composition calculation.

In the second phase, we will determine the percent composition of the salt mixture of $CaCl_2 \cdot 2H_2O$ and $K_2C_2O_4 \cdot H_2O$ by measuring the mass of the $CaC_2O_4 \cdot H_2O$ precipitate and using the supernatant (solution above the solid precipitate) to identify which reactant is limiting. In this phase, a simple test adding calcium chloride or potassium oxalate to the supernatant will tell us which reagent is present in excess. If calcium chloride is added to the supernatant and a precipitate forms, the potassium oxalate is present in excess as there is still potassium oxalate present in the supernatant to react with calcium chloride. If a precipitate is observed when potassium oxalate is added to the supernatant then we know that there is excess calcium chloride present to react with it and that the potassium oxalate is the limiting reagent. The sample problem below shows how to use these data to first find the mass of the limiting reactant and subsequently find the percent composition of the unknown mixture.

Sample Problem

A 1.00-g sample mixture of $CaCl_2 \cdot 2H_2O$ and $K_2C_2O_4 \cdot H_2O$ is added to water. After drying, the mass of the recovered $CaC_2O_4 \cdot H_2O$ precipitate is 0.354 g. Upon addition of $K_2C_2O_4 \cdot H_2O$ to the supernatant there is a small amount of precipitate formed indicating that $K_2C_2O_4 \cdot H_2O$ is the limiting reactant. What is the percent composition of the salt mixture? How many grams of $CaCl_2 \cdot 2H_2O$ were in the salt mixture?

Solution: To answer these questions we need the molar masses (MM) of the reactants and the product, and we need the balanced equation for the stoichiometric ratio of reactant to product. Using the stoichiometric ratios from eqn. 7-1 and eqn. 7-3, we see that the ratio of $K_2C_2O_4 \cdot H_2O$ to $CaC_2O_4 \cdot H_2O$ is 1:1. Because $K_2C_2O_4 \cdot H_2O$ is the limiting reactant we can use the stoichiometric ratio and the molecular weight of the product to determine the number of moles and grams of $K_2C_2O_4 \cdot H_2O$ using the following method:

Step A: Find the moles of $CaC_2O_4 \cdot H_2O$ product formed:

$$0.354 \text{ g } CaC_2O_4 \cdot H_2O \ / \ 146.12 \text{ g/mol} = 0.00242 \text{ mol } CaC_2O_4 \cdot H_2O.$$

Step B: Use the mole ratio to convert moles of product into moles of reactant:

$$0.00242 \text{ mol } CaC_2O_4 \cdot H_2O \ \times \ \frac{1 \text{ mol } K_2C_2O_4 \cdot H_2O}{1 \text{ mol } CaC_2O_4 \cdot H_2O} \ = \ 0.00242 \text{ mol } K_2C_2O_4 \cdot H_2O.$$

Step C: Convert moles of reactant to grams of reactant:

$$0.00242 \text{ mol } K_2C_2O_4 \cdot H_2O \ \times \ 184.24 \text{ g/mol} = 0.446 \text{ g } K_2C_2O_4 \cdot H_2O.$$

Therefore, the calculated mass of $K_2C_2O_4 \cdot H_2O$ in the original salt mixture is:

$$\frac{0.446 \text{ g } K_2C_2O_4 \cdot H_2O}{1.00 \text{ g mixture}} \ \times \ 100\% \ = \ 44.6\%.$$

This means that the mixture is made up of 44.6% $K_2C_2O_4 \cdot H_2O$ and 55.4% $CaCl_2 \cdot H_2O$ showing that $CaCl_2 \cdot 2H_2O$ was present in excess. Calculate the number of moles of $CaCl_2 \cdot 2H_2O$ present in the mixture.

PROCEDURE
Wear your safety goggles at all times in the laboratory.

For accuracy, it is recommended that the reaction be conducted in two trials. To expedite the process, you should conduct the two trials simultaneously. Obtain two 250-mL Erlenmeyer flasks or two 250-mL beakers from your instructor and weigh out 1.0 g of the $CaCl_2 \cdot 2H_2O$ and $K_2C_2O_4 \cdot H_2O$ mixture into each container. Label one container Trial 1 and the other container Trial 2 and follow the procedure below. Conducting the experiments simultaneously will be at the discretion of the lab instructor.

Part I. Calculating the mass of the precipitate
A. Formation of $CaC_2O_4 \cdot H_2O$ via reaction of $CaCl_2 \cdot 2H_2O$ with $K_2C_2O_4 \cdot H_2O$

1. Obtain and weigh a 250-mL Erlenmeyer flask. Record the mass in your report sheet.

2. Obtain 1.00 g of a salt mixture of $CaCl_2 \cdot 2H_2O/K_2C_2O_4 \cdot H_2O$ and add it to the 250-mL Erlenmeyer flask. Record the combined mass in your report sheet. NOTE: The % composition of this mixture is unknown.

3. Fill a 400-mL beaker with deionized water. Use pH paper to determine the acidity or basicity of the water. If the water is acidic, adjust it to be *just* basic with drops of 6 M NH_3. Acidic water will react with the oxalate ion and form oxalic acid. If the water is already neutral or basic *i.e.*, pH ≥ 7.0 it will not be necessary to adjust the pH.

4. Add ~150 mL of deionized water from step 3 to the Erlenmeyer flask containing the salt mixture from step 2. Stir the solution with a stirring rod for 2-3 minutes. The $CaC_2O_4 \cdot H_2O$ precipitate should begin to form. Let the precipitate settle before going on to the next step.

Repeat steps 1-4 for Trial 2.

B. Digestion of the precipitate.

1. Cover the Erlenmeyer flask containing the solution and $CaC_2O_4 \cdot H_2O$ precipitate with a watch glass and warm the solution on a hot plate. Use a thermometer to ensure that the temperature does not exceed 75°C. Heat the flask for 15 minutes with occasional stirring using your glass rod. Care should be taken not to remove precipitate as the overall yield will be affected. Any precipitate on the thermometer or glass rod should be rinsed back into the flask using the deionized water prepared in step A.4.

2. *After 15 minutes of heating,* remove the Erlenmeyer flask from the hot plate and allow the precipitate to settle; the solution does *not* need to cool to room temperature.

3. While the precipitate is settling, heat 50 mL of deionized water to 75°C for use as wash water in in the filtration step.

C. Filtration of the precipitate
NOTE: You should weigh your filter paper prior to filtration.

1. Set up a vacuum filtration apparatus by obtaining a Buchner funnel, a piece of filter paper, a vacuum flask and a piece of rubber tubing. It is important to seat the filter paper properly to prevent any solid from escaping around the sides of the paper. You can seal the filter paper by wetting it with 5mL of the deionized wash water prepared in B.3 above.

2. Once the precipitate has settled and the supernatant has cleared, obtain two 12 x 75-mm test tubes and label one $CaCl_2$ and the other $K_2C_2O_4$. With a pipette, remove 1 mL of supernatant and transfer it to the test tube labeled $CaCl_2$. Using the same pipette, remove 1 mL of supernatant and transfer it to the test tube labeled $K_2C_2O_4$. Save these samples for the limiting reactant test.

3. Filter the warm solution by slowly pouring it into the Buchner funnel. Be sure that the water is turned on and the hose is attached to your vacuum flask for proper suction. Use 5mL of the wash water prepared in step B3 to rinse any precipitate from the walls of the flask. You can repeat this step 2-3 times to ensure that all of the precipitate is recovered. Use a rubber policeman where necessary to dislodge any precipitate stuck to the sides of the flask.

4. Remove the filter paper and precipitate from the filter funnel. Air-dry the precipitate on the filter paper until the next laboratory period by placing it on a watch glass and allowing it to stand in your lab drawer until the following lab period. Alternately, you may use a drying oven set below 90°C to dry your sample. Allow the sample to dry for at least 1 hour. Once dry, determine the combined mass (±0.001 g) of the precipitate and filter paper and record it on your report sheet. After collecting the precipitate, you can move on to Part II.

Repeat this step for Trial 2.

Part II. Identifying the limiting reactant

Check with the lab instructor to determine if the samples collected in the two test tubes labeled "$CaCl_2$" and "$K_2C_2O_4$" need to be centrifuged.

A. Test for excess $K_2C_2O_4 \cdot H_2O$. Add two drops of 0.5 M $CaCl_2 \cdot 2H_2O$ to the supernatant liquid in the test tube labeled $CaCl_2$. If a precipitate forms, the $K_2C_2O_4 \cdot H_2O$ is *in excess* and $CaCl_2 \cdot 2H_2O$ is the limiting reactant in the original salt mixture.

B. Test for excess $CaCl_2 \cdot 2H_2O$. Add two drops of 0.5 M $K_2C_2O_4 \cdot H_2O$ to the supernatant liquid in test tube labeled $K_2C_2O_4$. If a precipitate forms, then $CaCl_2$ is *in excess* and $K_2C_2O_4 \cdot H_2O$ is the limiting reactant in the original salt mixture.

Repeat this step for Trial 2.

C. When you have completed two trials and are satisfied with the results obtained, rinse each beaker with small portions of warm water and discard in the Waste Liquids container. Then, rinse each beaker twice with tap water and twice with deionized water and discard in the sink.

FURTHER READING
1. https://fscimage.fishersci.com/msds/00211.htm (NH_3, last accessed July, 2019)
2. https://www.fishersci.com/shop/msdsproxy?productName=P273250&productDescription=POTASSIUM ($K_2C_2O_4$, last accessed July, 2019)
3. https://fscimage.fishersci.com/msds/21450.htm ($CaC_2O_4 \cdot H_2O$, last accessed July, 2019)
4. https://fscimage.fishersci.com/msds/03901.htm ($CaCl_2 \cdot 2H_2O$, last accessed July, 2019)
5. "A Conceptual Approach to Limiting-Reagent Problems" Sostarecz, M.C.; Sostarecz, A.G. *J. Chem. Educ.*, **89**, (2012), 1148.
6. "A Performance-Based Assessment for Limiting Reactants" Walker J.P.; Sampson, V.; Zimmerman, C.O.; Grooms, J.A. *J. Chem. Educ.*, **88**, (2011), 1243.
7. "A Nuts and Bolts Approach to Explain Limiting Reagents" Blankenship, C. *J. Chem. Educ.*, **64**, (1987), 13

COMMENTS

Name _____ Lab Section _____ Date_____ 99

Prelaboratory Assignment: Experiment 7
Limiting Reactants
Where appropriate, answers should be given to the correct number of significant digits.

1. The limiting reactant is determined in this experiment.
    a. What are the reactants (and their molar masses) in this experiment?

    b. How is the limiting reactant determined in the experiment?

2. Review the experimental procedure in Part B. What is the procedure and purpose of "digesting the precipitate"?

3. Two special steps in the procedure are incorporated to reduce the loss of the calcium oxalate precipitate. Identify the steps in the procedure and the reason for each step.

Experiment 7 Limiting Reactant

4. A 0.889-g sample of a CaCl2•2H2O/K2C2O4•H2O solid salt mixture is dissolved in ~ 150 mL of deionized water, previously adjusted to a pH that is basic. The precipitate, after having been filtered and air-dried, has a mass of 0.254 g. The limiting reactant in the salt mixture was later determined to be CaCl2•2H2O.

    a. Write the molecular form of the equation for the reaction.

    b. Write the net equation for the reaction.

    c. How many moles and grams of CaCl2•2H2O reacted in the reaction mixture?

    d. How many moles and grams of the excess reactant, $K_2C_2O_4 \cdot H_2O$, reacted in the mixture?

    e. How many grams of the $K_2C_2O_4 \cdot H_2O$ in the salt mixture remain unreacted (in excess)?

    f. What is the percent by mass of each salt in the mixture?

Name _____ Lab Section _____ Date _____ 101

REPORT ON EXPERIMENT 7
Limiting Reactant

Part I. Precipitation of $CaC_2O_4 \cdot H_2O$ from the Salt Mixture

	Trial 1	Trial 2
Unknown# _____		
Mass of Erlenmeyer flask	_____	_____
Mass of flask and salt mixture	_____	_____
Mass of salt mixture	_____	_____
Mass of filter paper	_____	_____
Mass of filter paper and $CaC_2O_4 \cdot H_2O$	_____	_____
Mass of air-dried $CaC_2O_4 \cdot H_2O$	_____	_____
Mass of oven dried $CaC_2O_4 \cdot H_2O$ (*Optional*)	_____	_____

Part II. Determination of Limiting Reactant

Write the molecular formula below

Limiting reactant in salt mixture _____

Excess reactant in salt mixture _____

Data Analysis

	Trial 1	Trial 2
Moles $CaC_2O_4 \cdot H_2O$ (or CaC_2O_4) Precipitated	_____	_____
Moles limiting reactant in salt mixture	_____	_____
Mass of limiting reactant in salt mixture	_____	_____
Mass of excess reactant in salt mixture	_____	_____
% limiting reactant in salt mixture	_____	_____
% excess reactant in salt mixture	_____	_____
Mass of excess reactant that reacted	_____	_____
Mass of excess reactant, unreacted	_____	_____

Experiment 7 Limiting Reactant

(Show all calculations for Trial 1. Submit your answers on a separate sheet as necessary.)
Where appropriate, answers should be given to the correct number of significant digits.

QUESTIONS

1. Without the digestion step, explain why the percent yield of the precipitate would decrease. How would this affect the calculation of the percent composition?

2. Calculate the mass of precipitate if the mass of the filter paper is 1.50 g and the combined mass of the precipitate and the filter paper is 3.28 g.

3. Why is it important to seal the filter paper to the funnel with a few drops of water prior to filtration?

4. A 2.00-g sample mixture of $CaCl_2 \cdot 2H_2O$ and $K_2C_2O_4 \cdot H_2O$ is added to water. After drying, the mass of the recovered $CaC_2O_4 \cdot H_2O$ precipitate is 0.654 g. Upon addition of $K_2C_2O_4 \cdot H_2O$ to the supernatant there is a small amount of precipitate formed indicating that $K_2C_2O_4 \cdot H_2O$ is the limiting reactant. What is the percent composition of the salt mixture?

5. Suppose that the $CaC_2O_4 \cdot H_2O$ in question 4 is mistakenly dried at 120°C rather than air dried. Calculate the percent composition of the salt mixture using the formula for anhydrous CaC_2O_4. What is the error introduced by this blunder?

Experiment 7 Limiting Reactant

Experiment 8
Standardization of a Sodium Hydroxide Solution

OBJECTIVE
To prepare and standardize a sodium hydroxide (NaOH) solution using a titration method; to determine the molar concentration of a strong acid.

EQUIPMENT
Part A: Three 125- or 250-mL Erlenmeyer flasks, 1000-mL beaker, 10-mL beaker, 150-mm test tube w/rubber-stopper, 500-mL polyethylene bottle, 50-mL buret, 3.5" funnel, Bunsen burner, weighing paper (or weigh boat), drying oven, desiccator, and an analytical balance.

Part B: Three 125- or 250-mL Erlenmeyer flasks, 25-mL pipet, pipet bulb, and a 50-mL buret.

REAGENTS
Part A: One liter boiled, deionized H_2O, 4 g NaOH, 2-3 g $KHC_8H_4O_4$, and 2 drops phenolphthalein. *Part B*: Three 10 mL samples of unknown acid solution (*i.e.*, HA or H_2A), 2 drops phenolphthalein, and variable amount of newly standardized NaOH solution.

SAFETY AND DISPOSAL
Refer to the MSDS information available online when working with NaOH,[1] $KHC_8H_4O_4$,[2] or phenolphthalein indicator solution.[3] Disposal for all compounds must be in accordance with local, state and federal regulations.

Disposal for $KHC_8H_4O_4$ should be into a labeled laboratory waste container for inorganic chemicals. Disposal for unknown acid solutions should be into a labeled laboratory waste container for acids. Disposal for NaOH should be into a labeled laboratory waste container for bases.

INTRODUCTION
A solution is standardized to determine the concentration of another solution of unknown concentration. In this lab, you will prepare and standardize a NaOH solution using a titration method. You will then use this standardized NaOH solution to help you determine the molar concentration of an unknown strong acid.

Among the properties of a *primary standard* are a high degree of purity, a relatively

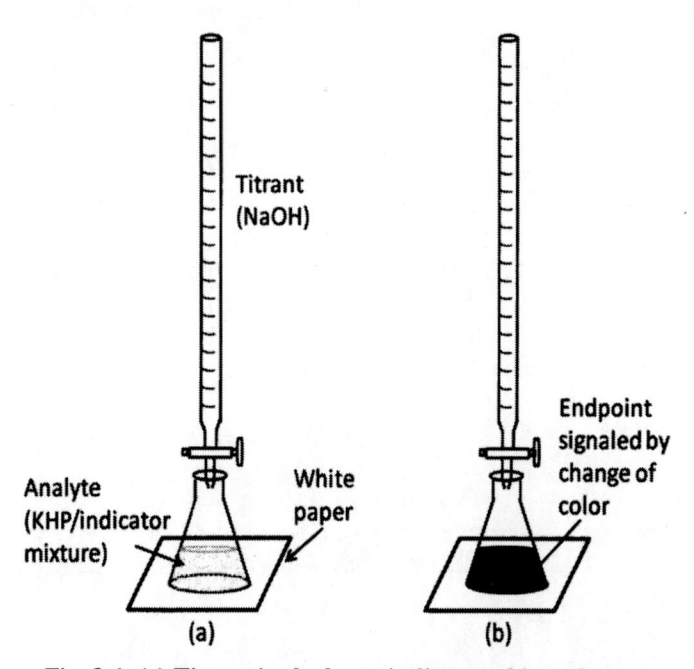

Fig. 8-1. (a) Titrant in the buret is dispensed into the flask until (b) the endpoint is reached

large molar mass, nonhygroscopic behavior, and reacting in a predictable way. Solid NaOH is *hygroscopic* (absorbs water vapor readily) and is thus not
considered a primary standard. To prepare an accurately known molar concentration of NaOH, it must be standardized with an acid that *is* a primary standard. In other words, it must be made into a ***standard solution***.

In Part A of this experiment, a titration will be used for the standardization process. This procedure requires a buret to dispense a liquid, called a ***titrant*** (in this case NaOH), into a flask containing the ***analyte*** (in this case $KHC_8H_4O_4$, potassium hydrogen phthalate; abbreviated KHP.) (See Fig. 8-1, part a). The buret is an accurately calibrated piece of glassware that will keep account of the volume of base (NaOH) used to react with the acid (KHP). Since KHP (see Fig. 8-2) can be weighed accurately, the NaOH concentration can be reliably determined from the volume of the base and the mass of the acid used in the titration.

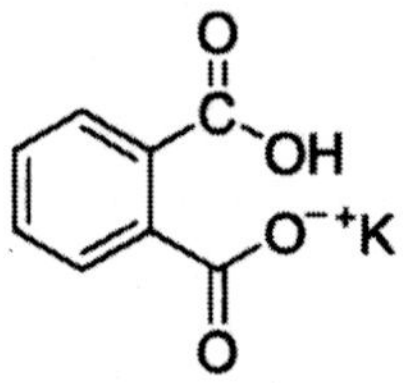

Fig. 8-2. Structure of KHP

After weighing, the KHP is transferred to an Erlenmeyer flask and dissolved in deionized H_2O. A small amount of the indicator phenolphthalein is added. NaOH solution is then dispensed into the KHP/phenolphthalein mixture from a buret. As NaOH is added, it reacts with KHP according to the following reaction:

$$KHC_8H_4O_4(aq) + NaOH(aq) \rightarrow H_2O(l) + NaKC_8H_4O_4(aq) \quad (8\text{-}1)$$

$$HC_8H_4O_4^-(aq) + OH^-(aq) \rightarrow H_2O(l) + C_8H_4O_4^{2-}(aq) \quad (8\text{-}2)$$

NaOH is added until the ***equivalence point*** is reached, when the moles of NaOH added are exactly equal to the moles of KHP present in the flask. Unfortunately, it is impossible for the human eye to detect the equivalence point! Commonly, we detect the ***endpoint***, which is the point where the indicator being used changes color. Phenolphthalein, which is colorless in acid and pink in base, signals the end of the reaction (see Fig. 8-1, part b). Just one drop of excess NaOH results in a color change of phenolphthalein from colorless to pink. At this point the titration has ended and the total volume of NaOH delivered is determined. The NaOH solution molar concentration can now be calculated from the volume of NaOH delivered and the mass of KHP used.

Note: *Care must be taken to minimize KHP exposure to atmospheric moisture as it is mildly hygroscopic and will slightly absorb water, leading to inaccuracies in the mass measurement. KHP might be supplied to you as an oven-dried sample (ask your instructor).*

In Part B, an unknown molar concentration of a strong acid solution is determined. You will use the standardized NaOH solution (*now with a known molar concentration!*) you prepared in Part A to help you determine the concentration of the unknown strong acid. By knowing the volume and molar concentration of the NaOH, the moles of the unknown strong acid neutralized in the reaction can be calculated. An acid-base reaction like the one performed here is also called a ***neutralization reaction*** with generic formula:

$$\text{a base } + \text{ an acid } \rightarrow \text{ water } + \text{ a salt} \quad (8\text{-}3)$$

If your unknown strong acid is monoprotic, HA (*e.g.*, HCl), then the mole ratio of acid to NaOH is 1:1. However, if your acid is diprotic, H_2A (*e.g.*, H_2SO_4), then the mole ratio of strong acid to NaOH is 1:2. Your instructor will inform you of the acid type.

Experiment 8 Standardization of a Sodium Hydroxide Solution

PROCEDURE
Wear your safety goggles at all times in the laboratory.

Part A: Standardization of a Sodium Hydroxide Solution.

You will need to prepare approximately one liter of boiled, deionized H_2O for this experiment. Boiling the H_2O removes traces of CO_2 that would react with the NaOH solution to form a less soluble salt, sodium carbonate (Na_2CO_3). You are to complete at least three good ($\pm$ 1% reproducibility) standardization trials. Prepare and clean three 125- or 250-mL Erlenmeyer flasks for the titration.

1. Prepare boiled, deionized H_2O. Pour ~1 L of deionized H_2O into 1000-mL beaker and place on a Bunsen burner until boiling; cool to room temperature.
2. Prepare stock NaOH solution.
    a. (***Caution***: *NaOH is extremely corrosive-do not allow skin contact. Any spills to skin, eyes, or clothing should be flushed thoroughly with H_2O.*) Dissolve ~4 g of NaOH in 15 mL of deionized H_2O in a 150-mm rubber-stoppered test tube.
    b. Mix thoroughly. Allow solution to stand for the precipitation of sodium carbonate.
3. Dry the primary standard acid. Place 2-3 g of KHP in a 10-mL beaker and dry at 110 °C for several hours in a drying oven. Cool the sample in a desiccator.
4. Prepare diluted NaOH solution.
    a. Decant ~250 mL of previously boiled, deionized H_2O into a 500-mL polyethylene bottle.
    b. Decant ~4 mL of NaOH solution prepared in part A.1 into the 500-mL polyethylene bottle.
    c. Dilute to 500 mL with previously boiled, deionized H_2O.
    d. Cap the polyethylene bottle and swirl solution. Label the bottle. Calculate an *approximate* molar concentration of your diluted NaOH solution.
5. Prepare primary standard acid.
    a. Calculate the mass of KHP that will require ~15-20 mL of your diluted NaOH solution to reach the equivalence point. Show calculations on your lab assignment.
    b. Measure calculated mass ($\pm$ 0.001 g) of KHP on a tared piece of weighing paper. Transfer to a clean, labeled Erlenmeyer flask. Similarly, prepare all three samples while you are occupying the balance.
    c. Dissolve KHP in ~50 mL of previously boiled, deionized H_2O and add 2 drops of phenolphthalein.
6. Prepare and fill buret.
    a. Wash 50-mL buret and funnel thoroughly with soap and H_2O. Flush with tap H_2O and rinse several times with deionized H_2O.
    b. Rinse buret three times with 5 mL portions of diluted NaOH solution, making certain the solution wets the entire inner surface. Discard each rinse in the Waste Bases container. Have your instructor approve your buret before continuing.
    c. Using the cleaned funnel, fill buret with NaOH solution making sure to remove all bubbles from the buret tip.
    d. After 10-15 seconds, read the volume by viewing the bottom of the meniscus.
    e. Record initial volume using the correct number of significant figures.
    f. Place sheet of white paper beneath the Erlenmeyer flask.

7. Titrate primary standard acid.
    a. Slowly add the NaOH titrant to the first acid sample. Swirl flask after each addition. Initially, add NaOH in 1-2 mL increments.
    b. As you get closer to the endpoint the pink color will persist for longer and longer periods of time before disappearing. Continue adding NaOH more slowly at this point (one drop at a time).

c. Continue addition of NaOH until endpoint is reached. The light pink color should not disappear. DO NOT ADD more NaOH than necessary to reach this constant color.
   d. Record final volume of NaOH in buret.

8. Repeat analysis with remaining acid samples. Refill the buret with NaOH and repeat titration.
9. Calculations. Calculate the average molar concentration of the NaOH solution. Molarities should be within ± 1%. Place a new label on polyethylene bottle.
10. Disposal: Discard the neutralized solutions in the Waste Acids container.

Part B: Molar Concentration of an Acid Solution.
Three samples of unknown acid will be analyzed. Thus, prepare three clean 125 or 250-mL Erlenmeyer flasks.
1. Prepare acid samples of unknown concentration. Pipet 10.00 mL of the unknown acid solution into an Erlenmeyer flask. Add 2 drops of phenolphthalein.
2. Fill buret and titrate.
   a. Refill buret with the (now) standardized NaOH solution. Record the initial volume.
   b. Titrate the unknown acid sample until endpoint is reached. Refer to Part A.7.
   c. Record final volume of NaOH in buret.
3. Repeat. Similarly titrate the remaining unknown acid samples.
4. Calculations. Calculate the average molar concentration of the unknown acid.
5. Disposal. Discard the neutralized solutions in the Waste Acids container.
6. Cleanup. Rinse buret and pipet several times with tap H_2O and discard through the tip into sink. Rinse twice with deionized H_2O. Similarly clean the Erlenmeyer flasks. All solids should be discarded in the Waste Solid Acids container.

SAVE: Save your standardized NaOH solution in the tightly capped 500-mL bottle for Experiment 9 and your boiled, deionized water in a wash bottle.

FURTHER READING
1. http://www.labchem.com/tools/msds/msds/LC23900.pdf (NaOH, last accessed July, 2019).
2. https://fscimage.fishersci.com/msds/19425.htm ($KHC_8H_4O_4$, last accessed July, 2019).
3. https://fscimage.fishersci.com/msds/96382.htm (phenolphthalein, last accessed July, 2019).
4. Kumar, V., P. Courie, and S. Haley, "Quantitative Microscale Determination of Vitamin-C" *J. Chem. Educ.,* **69**(8) (1992) A213-A214.
5. McAlpine, R.K., "The carbon dioxide problem in neutralization titrations." *J. Chem. Educ.,* **21**(12) (1944). 589.
6. Smith, B.W. and M.L. Parsons, "Preparation of standard solutions." *J. Chem. Educ.,* **50**(10), (1973) 679.

Prelaboratory Assignment: Experiment 8
Standardization of a Sodium Hydroxide Solution
Where appropriate, answers should be given to the correct number of significant digits.

1. What is the difference between a solution and a standard solution?

2. Define the analyte in a titration.

3. What is the primary standard used in this experiment (name and formula)? Define primary standard.

4. What is the difference between a primary standard and a secondary standard?

5. Distinguish between an equivalence point and an endpoint.

6. Explain the difference between equations 8-1 and 8-2

Experiment 8 Standardization of a Sodium Hydroxide Solution

7. In the procedure, a 4.00 g mass of NaOH is dissolved in 5.00 mL of H_2O.
   a. What is the approximate molar concentration of the NaOH?

   b. In the procedure, a 4.00 mL aliquot of this solution is diluted to 5.00×10^2 mL. What is the approximate molar concentration of NaOH in the diluted solution? Record answer here as well as in lab assignment.

8. Calculate the mass of KHP (molar mass 204.44 g/mol) that reacts with 15 mL of the NaOH solution in the procedure. Express mass of KHP with the correct number of significant figures. Record answer in lab assignment.

9. a. A 0.411 g sample of KHP is dissolved with 50.00 mL of deionized H_2O in a 125-mL Erlenmeyer flask. The sample is titrated to the endpoint with 15.17 mL of a NaOH solution. What is the molar concentration of the NaOH solution?
   Give answer to the correct number of significant figures.

   b. A 25.00 mL aliquot of nitric acid (HNO_3) of unknown concentration is pipetted into a 125-mL Erlenmeyer flask and 2 drops of phenolphthalein are added. The *above* NaOH (titrant) solution is used to titrate the HNO_3 (analyte). If 16.77 mL of titrant is used to reach the endpoint, what is the molar concentration of the nitric acid? Give answer to the correct number of significant figures.

Experiment 8 Standardization of a Sodium Hydroxide Solution

Name _____ Lab Section _____ Date _____ 109

REPORT ON EXPERIMENT 8
Standardization of a Sodium Hydroxide Solution

Part A

Approximate concentration of NaOH solution (Part A.3, calculated in pre-lab assignment) _____

Approximate mass of KHP needed (Part A.4; calculated in pre-lab assignment) _____

	Trial 1	Trial 2	Trial 3
Tared mass of KHP (*g*)			
Moles of KHP (*mol*)			
Initial buret reading of NaOH (mL)			
Final buret reading of NaOH (mL)			
Volume of NaOH dispensed (mL)			
Molar concentration of NaOH (*M*)			
Average molar concentration of NaOH (*M*)			
Standard deviation of molar concentration			
Relative standard deviation of molar concentration (*%RSD*)			

Part B

Acid type: _____   Unknown No. _____

	Sample 1	Sample 2	Sample 3
Volume of acid solution (mL)	10.0	10.0	10.0
Initial buret reading of NaOH (mL)			
Final buret reading of NaOH (mL)			
Volume of NaOH dispensed (mL)			
Molar concentration of NaOH (*M*), Part A			
Moles of NaOH dispensed			
Molar concentration of acid solution (*M*)			
Average molar concentration of acid solution (*M*)			
Standard deviation of molar concentration			
Relative standard deviation of molar concentration (*%RSD*)			

Experiment 8 Standardization of a Sodium Hydroxide Solution

QUESTIONS *(Submit your answers on a separate sheet as necessary.)*
Where appropriate, answers should be given to the correct number of significant digits.

1. A 0.5112-g sample of 99.99% pure KHP (M.W. 204.44) was dissolved in about 25 mL of distilled water and titrated to the phenolphthalein end point with 26.93 mL of a sodium hydroxide solution. Calculate the molar concentration of the sodium hydroxide solution.

2. A student obtains the following concentrations for her sodium hydroxide solution from three individual titrations: 0.1000 M, 0.0998 M, 0.1004 M. Calculate the average, standard deviation, and %RSD for these results.

3. What are the (at least) three important characteristics of a primary standard?

4. Describe the results you obtained. What was the largest source of error (uncertainty) in your measurements? What will you do to improve your results for the next time you do a titration?

Experiment 8 Standardization of a Sodium Hydroxide Solution

Experiment 9
Volumetric Analysis of Vinegar

OBJECTIVE
To determine the percent by mass of acetic acid in vinegar using a volumetric titration.

EQUIPMENT
Two 125- or 250-mL Erlenmeyer flasks, two 10-mL graduated cylinders, 50-mL buret, 3.5"-funnel, one small sheet of white paper, and an analytical balance.

REAGENTS
150 mL of standardized NaOH (from Experiment 8), 10 mL each of two unknown vinegars, 20 mL of boiled, deionized H_2O (from Experiment 8), and 2 drops phenolphthalein indicator solution.

SAFETY AND DISPOSAL
Refer to the MSDS information available online when working with NaOH,[1] CH_3COOH,[2] or phenolphthalein indicator solution.[3]

Disposal for all compounds must be in accordance with local, state and federal regulations. Disposal for unknown acid solutions should be into a labeled laboratory waste container for acids. Disposal for NaOH should be into a labeled laboratory waste container for bases.

INTRODUCTION

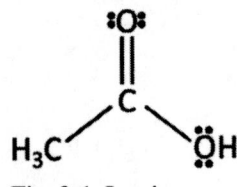

Fig. 9-1. Lewis structure of acetic acid

Use standardized NaOH to determine the percent by mass of acetic acid in vinegar. Household vinegar is ~5% (by mass) acetic acid, CH_3COOH (see Fig. 9-1). Volumetric analysis of CH_3COOH involves calculating a known mass of vinegar and titrating with a measured volume of the NaOH solution. Since the volume and molar concentration of the standardized NaOH are known, the number of moles of NaOH used for the analysis are also known. In order to calculate the percent by mass of acetic acid you must know the stoichiometry of reaction between CH_3COOH and NaOH. From the balanced equation:

$$CH_3COOH(aq) + NaOH(aq) \rightarrow NaCH_3CO_2(aq) + H_2O(l)$$

The number of moles of CH_3COOH reactant are equal to the number of moles of NaOH reactant. The mass of CH_3COOH is calculated from the neutralization reaction using its molar mass, 60.05 g/mol:

$$(\text{mol } CH_3COOH) \times (60.05 \text{ g } CH_3COOH/\text{mol } CH_3COOH) = \text{g } CH_3COOH$$

The percent by mass of CH_3COOH in vinegar is then easily calculated:

$$(\text{g } CH_3COOH / \text{g vinegar}) \times 100\% = \% \text{ by mass } CH_3COOH$$

PROCEDURE
Wear your safety goggles at all times in the laboratory.

Preparation and Analysis of Vinegar Sample.
Samples of two vinegars are analyzed for the amount of CH_3COOH in each sample. The standardized NaOH solution and boiled water prepared in Experiment 8 are to be used for this experiment. If these solutions were not saved you must either prepare them again or obtain already prepared solutions from your instructor. Obtain 10 mL of each of two vinegars in separate 10-mL graduated cylinders. Clean four 125- or 250-mL Erlenmeyer flasks for the titration.

1. Calculate the volume of vinegar. Calculate the volume of vinegar that would be needed for the neutralization of 25.0 mL of the standardized NaOH solution. Assume the vinegar has a density of 1.00 g/mL, a percent acetic acid of 5.00% by mass, and the standardized NaOH is 0.100 M NaOH. Show the calculation on your laboratory assignment
2. Prepare the vinegar sample.
    i. Record the mass of a clean and dry 125- or 250- mL Erlenmeyer flask ($\pm$ 0.01 g).
    ii. Add the calculated volume of one brand of vinegar to the flask. Record mass.
    iii. Add 2 drops of phenolphthalein solution.
    iv. Rinse wall of flask with 20 mL of previously boiled, deionized H_2O.
3. Prepare buret and titration setup.
    i. Rinse twice a clean 50-mL buret with ~5 mL of standardized NaOH, making certain no drops cling to the inside wall.
    ii. Fill buret with standardized NaOH. Eliminate all air bubbles from the buret tip. After 10-15 seconds, record the initial volume.
    iii. Place a sheet of white paper beneath the flask.
4. Titrate vinegar sample.
    i. Record the exact molar concentration that you previously calculated for your standardized NaOH (from Experiment 8).
    ii. Slowly add NaOH from the buret to the flask, swirling after each addition.
    iii. Occasionally, rinse the wall of flask with previously boiled, deionized water.
    iv. Continue addition of NaOH titrant until endpoint is reached. After 10-15 seconds, read and record final volume of NaOH left in the buret.
5. Repeat with same vinegar. Refill the buret and repeat the titration at least once more with another sample of the *same* vinegar.
6. Repeat with different vinegar. Repeat Parts 1-5 for a second vinegar sample to determine its average % CH_3COOH.
7. Calculations. Determine the average percent by mass of CH_3COOH in the vinegar(s).
8. Cleanup. Rinse buret several times with tap H_2O and discard through the tip into sink. Rinse twice with deionized H_2O. Similarly clean the Erlenmeyer flasks.

FURTHER READING
1. http://www.labchem.com/tools/msds/msds/LC24350.pdf (NaOH, last accessed July, 2019).
2. http://www.labchem.com/tools/msds/msds/LC10100.pdf (CH_3COOH, last accessed July, 2019).
3. https://fscimage.fishersci.com/msds/96382.htm (phenolphthalein, last accessed July, 2019).
4. Castillo, C. A., Jaramillo, A., "An alternative procedure for titration curves of a mixture of acids of different strengths." *J. Chem. Educ.*, **66**(4), (1989), 341.
5. McMills, L., Nyasulu, F., Barlag, R., "Comparing Mass and Volumetric Titrations in the General Chemistry Laboratory." *J. Chem. Educ.*, **89**(7), (2012), 958-959.

Name _____ Lab Section _____ Date _____ 113

Prelaboratory Assignment: Experiment 9
Volumetric Analysis of Vinegar

Where appropriate, answers should be given to the correct number of significant digits.

1. If it takes 25.00 mL of 0.0500 M HCl to neutralize 15.00 mL of NaOH solution, what is the concentration of the NaOH solution?

2. A 21.50 mL volume of 0.0950 M NaOH is required to reach the phenolphthalein endpoint in the titration of a 3.15 g sample of vinegar.
    a. Calculate the number of moles of CH_3COOH in the vinegar sample.

    b. Calculate the mass of CH_3COOH in the vinegar sample.

    c. Calculate the percent by mass CH_3COOH in the vinegar sample. Assume the density of the vinegar is 1.00 g/mL.

Experiment 10 Determination of the Molar Gas Constant, R

3. A vinegar sample contains 2.00 × 10⁻³ moles of acetic acid. If a 4.00 g sample of vinegar was used in the titration, calculate the percent by mass of this vinegar solution. Answer should be given to the correct number of significant figures.

4. Write a balanced net ionic equation for the reaction occurring in this experiment.

5. One of the characteristics of acids is that they taste sour. Can you think of any foods or edible materials that have a sour taste? Make a list of these foods and substances

6. The pH of the gastric juices in your stomach is ~1.4. If you titrated a sample of gastric juice as you did the vinegar, would it take more or less NaOH to neutralize the gastric juice than the vinegar? Explain your reasoning.

Experiment 10 Determination of the Molar Gas Constant, R

Name _____ Lab Section _____ Date _____ 115

REPORT ON EXPERIMENT 9
Volumetric Analysis of Vinegar
Where appropriate, answers should be given to the correct number of significant digits.

Calculate the approximate percent by mass of acetic acid in the vinegar sample needed for the analysis.

Brand of vinegar or unknown number: _____   _____

	Trial 1	Trial 2	Trial 1	Trial 2
Mass of flask (g)				
Mass of flask + vinegar (g)				
Mass of vinegar (g)				
Initial buret reading of NaOH (mL)				
Final buret reading of NaOH (mL)				
Volume of NaOH used (mL)				
Molar concentration of NaOH (M)				
Moles of NaOH added (mol)				
Moles of CH_3COOH in vinegar (mol)				
Mass of CH_3COOH in vinegar (g)				
Percent by mass of CH_3COOH in vinegar (%)				
Average percent by mass of CH_3COOH in vinegar (%)				

Calculations.

Briefly compare the two vinegars with respect to mass percent, odor, and clarity.

Experiment 10 Determination of the Molar Gas Constant, R

QUESTIONS

(Submit your answers on a separate sheet as necessary.)

1. What is the purpose of the indicator? How does it tell you when the titration is complete?

2. Give two reasons for the statement below: (*Hint: Consider both your own experiment and that of the person who uses the balance after you*)
   Always dry the outside surfaces of beakers, flasks, and other pieces of glassware before you place them on the balance pan.

3. Fill in the blanks for the following by circling the answer for each missing word:

   a. Phenolphthalein is: _____ in acidic solutions,   colorless pink

   _____ in neutral solutions,   colorless pink

   _____ in basic solutions.   colorless pink

   b. Acidic solutions contain _____ ions, and H^+ OH^-

   basic solutions contain _____ ions. H^+ OH^-

4. Explain why the solution being titrated first turns pink then goes colorless before the endpoint is reached.

5. A student gets a dark pink color for trial 2 but uses that data to calculate the mass percent of acetic acid. Would the mass percent be too high or too low? Explain and be specific

Experiment 10 Determination of the Molar Gas Constant, R

Experiment 10
Determination of the Molar Gas Constant, R

OBJECTIVE
To determine the molar gas constant, R, by measuring the volume, pressure, and temperature of a known mass of oxygen collected over water.

EQUIPMENT
0-100°C thermometer, 70-cm length of 4-mm tubing, ring stand with 3-in. iron ring, Bunsen burner with wing top, spatula, pinch clamp, 15-cm length of 1/4-in. rubber tubing, one-hole no. 2 rubber stopper, two-hole no. 6 rubber stopper, 600-mL beaker, 50-mL graduated cylinder, 500-mL round bottom flask, 18 x 150-mm test tube, evaporating dish, crucible.

REAGENTS
3 g of $KClO_3$, 0.2 g of MnO_2

SAFETY AND DISPOSAL
Refer to the MSDS information available online when working with $KClO_3$,[1] MnO_2,[2] and KCl.[3] Disposal for all compounds must be in accordance with local, state and federal regulations. Disposal for KCl and MnO_2 should be into a labeled laboratory waste container for inorganic chemicals.

INTRODUCTION
The volume, V, for a given mass of an ideal gas is directly proportional to its absolute temperature, T, and inversely proportional to its pressure, P. If n is the number of moles of gas, the relationship of volume, pressure and the number of moles is given by the ideal gas law,

$$PV = nRT$$

where R is a proportionality constant. Solving for R,

$$R = \frac{PV}{nT} \tag{10-1}$$

R is known as the molar gas constant. R has dimensions of energy per Kelvin per mole, and its numerical value depends on the units used to express P and V. Thus, if P is in atmospheres and V is in liters, R will be expressed in liter-atmospheres per Kelvin per mole.

In this experiment, the value of R will be determined by measurement of P, V, T, and n on a sample of oxygen gas which will be produced by the reaction

$$2KClO_3\ (s) \rightarrow 2KCl\ (s) + 3O_2\ (g)$$

The physical properties of oxygen gas, when measured at room temperature, closely resemble those of an ideal gas. There is a loss of material in the reaction vessel as the reaction proceeds, because oxygen is liberated as a gas. The number of moles of oxygen liberated, *n*, can be expressed as

$$n = \frac{\text{loss of mass of material in a reaction vessel}}{\text{molecular weight of oxygen}}$$

The volume, *V*, occupied by the sample of oxygen produced in the reaction is measured by the displacement of water. For each milliliter of O_2 liberated in the reaction, 1 mL of water is displaced. The pressure, *P*, of the sample will equal the atmospheric pressure corrected for the vapor pressure of water. The correction is needed because the sample of oxygen that is collected is saturated with water vapor. Finally, the temperature, *T*, of the gas is measured with a thermometer when the system is at equilibrium with its surroundings. When *n*, *V*, *P*, and *T* are all known, *R* may be calculated using eqn. 10-1.

PROCEDURE

Wear your safety goggles. Clean a test tube (18 × 150-mm) and set it in an oven to dry. In the meantime, construct the items shown in Fig. 10-1 from lengths of 4-mm glass tubing. Be sure to fire-polish the ends of all glass tubing, except the capillary tip. Review the material on handling glass tubing and stoppers in the Laboratory Safety section and then assemble the apparatus shown in Fig. 10-2. Use a few drops of glycerol or water to lubricate rubber tubing or the holes of stoppers before you insert glass tubing into them. Be sure that tube C extends a few mm above the bottom of the 500-mL flask and that tubes A and B protrude slightly below the stoppers. Wipe clean of excess glycerol both the glass tubing and the inside of the test tube. *Do not let the glycerol lubricant come in contact with the $KClO_3$. The resultant mixture is potentially explosive.*

Test the assembly for leaks by the following procedure. Fill the reservoir flask (F) to the base of the neck with water and return it to the assembled apparatus. Remove the test tube and blow through glass tube A so that enough air enters the flask to raise the level of the water in tube C an inch or two above the water level of the flask. Immediately close off the system by bending back rubber tubing X. If the raised level of water in tube C does not remain steady, there is a leak in the system. Tighten all connections or replace suspected parts (particularly the pinch clamp) and test again.

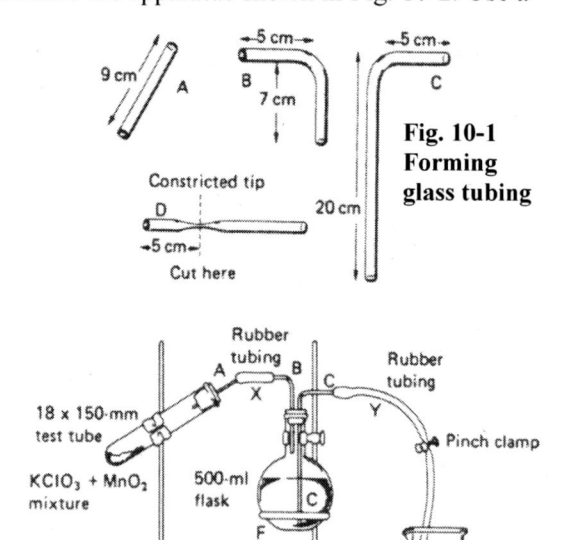

Fig. 10-1 Forming glass tubing

Fig. 10-2. Assembled apparatus

Carefully dry about 3 g of $KClO_3$ by gently warming it in an evaporating dish above a small flame for about 2 minutes. Similarly, dry about 0.2 g of MnO_2 in a clean crucible. Then, with a spatula, carefully mix the MnO_2 with the $KClO_3$ in the evaporating dish.

Remove the dry test tube from the oven. When it has cooled to room temperature, pour the MnO_2-$KClO_3$ mixture into it through a paper funnel. Take care that none of the mixture is left on the upper part of the test tube where it might come into contact with the rubber stopper.

Determine the mass of the test tube and its contents on an analytical balance and record the mass on your data sheet.

Open the pinch clamp and blow through glass tube A so that tube C, the rubber tube Y, and the constricted glass tube (D) are completely filled with water. Close the pinch clamp. Now attach the test tube containing the MnO_2-$KClO_3$ mixture to the apparatus. Carefully open the pinch clamp. At this point a little water may flow out of the constricted tip. However, a continuous siphoning of water indicates a leak. If such is the case, close the pinch clamp, tighten all connections, and repeat the procedure.

Pour about 50 mL of water into the 600-mL beaker. Make sure the constricted tip is below the surface of the water, open the pinch clamp, and then raise the beaker until the water levels in the beaker and in the reservoir flask are the same. Still keeping the levels equal, close the pinch clamp. The pressure of the air within the apparatus is now equal to the barometric pressure. Discard the water in the beaker and put the beaker back under the constricted tip. Then open the pinch clamp and *leave it open.*

Have your instructor check the apparatus before you proceed. Cautiously heat the MnO_2-$KClO_3$ mixture by moving the flame back and forth across the test tube so that the evolved oxygen gas displaces the water from the flask at a steady rate. The mixture will gently bubble and turn dark. Throughout this procedure the constricted tip must remain under the water which has been displaced by the oxygen. (The appearance of a white fog in the test tube or splattering of the contents indicates that heat is being applied too strongly -- use a steady *low* flame.) Do *not* allow the water level in flask (F) to fall below the tip of tube (C).

When 200- 250 mL of water has been collected in the beaker, stop heating. *Allow the test tube to cool to room temperature with the pinch clamp remaining open and the tip (D) under water.* Then equalize the water levels in the beaker and in the flasks as before, and close the clamp.

In a graduated cylinder, carefully measure the volume of water collected in the beaker. (Note 10-1). Express this volume in liters and record it on your data sheet as the volume of evolved oxygen gas. Measure the temperature of the water and record it on the data sheet as the temperature of the evolved oxygen. Detach the test tube, taking care not to spill any of its contents, weigh it on the analytical balance, and record its mass. Record the barometric pressure (this value will be provided by the laboratory instructor). See Appendix B, Table 2 for the vapor pressure of water at the recorded temperature and subtract this value from the barometric pressure to obtain the pressure of the "dry" oxygen.

From the loss in mass of the test tube and its contents, calculate the number of moles of oxygen liberated. Calculate the volume per mole and the gas constant, R.

Optional calculations. Report the results of the determination of the gas constant, R, to your instructor, who will list the results of the entire class. Copy these data onto the report sheet. Review the material on uncertainty and error analysis in the Mathematical Treatment of Data section and compute:
1. the mean of the results
2. the deviation of your value from the mean
3. the standard deviation
4. the 95% confidence level

Show the details of all calculations and complete all items asked for in the report sheet.

NOTES
10-1. The volume of the displaced water may be determined alternately from its mass and density. Weigh the beaker and water on a platform balance and then weigh the empty beaker. See Appendix B, Table 3 for the water density at the recorded temperature.

FURTHER READING
1. https://fscimage.fishersci.com/msds/19300.htm ($KClO_3$, last accessed July, 2019)
2. https://fscimage.fishersci.com/msds/13610.htm (MnO_2, last accessed July, 2019)
3. http://www.labchem.com/tools/msds/msds/LC18790.pdf (KCl, last accessed July, 2019)
4. Bastrop, O., Demandt, K, and Hansen, KO., "The Thermal Decomposition of $KClO_3$, Textbook Errors.", *J. Chem. Educ.,* **39** (1962) 573.

COMMENTS

Name _____ Lab Section _____ Date _____

Prelaboratory Assignment: Experiment 10
Determination of the Molar Gas Constant *R*

Where appropriate, answers should be given to the correct number of significant digits.

1. Which of the following measurements determines the number of significant digits used to express the value of the gas constant, *R*: pressure, temperature, loss of mass, or volume? Explain your answer fully.

2. In this experiment, is it necessary to completely decompose all of the $KClO_3$? Explain your answer fully.

3. Why do you need to "take care that none of the $KClO_3 - MnO_2$ mixture is left on the upper part of the test tube where it might come in contact with the rubber stopper?" Explain your answer fully.

Experiment 10 Determination of the Molar Gas Constant, R

4. Indicate whether each of the following experimental "blunders" would produce a high result, a low result, or make no difference in the value obtained for the volume of gas per mole, V/n.

   a. The $KClO_3$ was not dried before use.

   b. The $KClO_3$ lost oxygen during the preliminary drying step.

   c. Air leaked into the apparatus during heating.

   d. In your calculations you failed to correct for the vapor pressure of water.

   e. Measurement of the volume of water without waiting for the apparatus to cool.

Name _____ Lab Section _____ Date _____

REPORT ON EXPERIMENT 10
Determination of the Molar Gas Constant, *R*

Where appropriate, answers should be given to the correct number of significant digits.

DATA AND RESULTS

mass of test tube, $KClO_3$, and MnO_2 catalyst _____

mass of test tube and residue _____

mass of oxygen evolved _____

moles of oxygen evolved, n _____

temperature of oxygen, T _____

barometric pressure _____

vapor pressure of water at T _____

pressure of "dry" oxygen _____

volume of water collected=volume of oxygen _____

volume per mole of oxygen, V/n _____

gas constant, R _____

Optional calculations: class-determined values of R

_____ _____ _____ _____ _____ _____ _____ _____

_____ _____ _____ _____ _____ _____ _____ _____

_____ _____ _____ _____ _____ _____ _____ _____

mean value _____

deviation _____

standard deviation _____

95% confidence level _____

Show the details of all calculations; use extra sheets if necessary.

Experiment 10 Determination of the Molar Gas Constant, R

QUESTIONS

(Submit your answers on separate sheet if necessary.)

1. Does the pressure of air initially in the flask introduce an error? Explain your answer.

2. Is it necessary to consider the pressure of water vapor initially in the flask? Explain your answer.

3. Briefly discuss whether each of the following "mistakes" would produce a high result, a low result, or make no difference in the value obtained for the gas constant, R.
    i. The water level in the flask dropped below the tip of tube C.

    ii. Loss of some potassium chlorate from the test tube.

    iii. Incomplete decomposition of the potassium chlorate.

    iv. The correction for the vapor pressure of water was omitted in the calculation.

    v. Failure to use the MnO_2 catalyst.

4. Why is it necessary to equalize the pressure of the receiving beaker and the flask before measuring the delivered volume of water?

Experiment 11
Recognizing the "Fingerprints" of Atoms and Molecules

OBJECTIVES
To differentiate between the continuous spectra of black body radiators and atomic line spectra. To calculate lines in the Rydberg series and recognize the Balmer series of hydrogen atom. To recognize that different atoms will have different emission spectra. To differentiate between absorption and emission of atoms and molecules. To define the difference between short-lived fluorescence and long-lived phosphorescence.

EQUIPMENT
Spectroscopes, calibrated spectroscopes, gas emission tubes (H_2, He, Ne), gas emission power supply, incandescent light bulb and lamp, UV lamp, visible absorption spectrometer (Note 11-1), light sticks or glow sticks, solution of green food coloring, glow-in-the-dark toys, 1-cm cuvettes.

REAGENTS
Solutions of red and yellow commercial fruit drink, diluted to maximum absorbance of 1.

SAFETY AND DISPOSAL
Do not look directly into the sun or other intense light source such as a UV lamp. Dispose of used light sticks as per manufacturer's instructions.

INTRODUCTION
Light is one of the premiere tools in science for probing a system to determine its physical state, molecular structure, or even its identity. The electromagnetic spectrum is divided up into regions based upon wavelength. Wavelength λ (the Greek letter lambda) is directly related to frequency ν (the Greek letter nu) by Eq. 11-1:

$$\lambda = c/\nu \qquad (11\text{-}1)$$

where c is the speed of light, 3.00×10^8 m/s. In addition, light energy E is directly related to wavelength as shown in Eq. 11-2:

$$E = hc/\lambda = h\nu \qquad (11\text{-}2)$$

where h is Planck's constant, 6.626×10^{-34} J•s. Because of these two relationships, the wavelength of light automatically indicates a particular frequency and energy for that light.

The typical breakdown of the electromagnetic spectrum is given in Table 11-1.

Type	Wavelength (nm)	Frequency (s^{-1} or Hz)	Energy (J)	Interaction with matter
Gamma	<0.1	>10^{20}	10^{-14}	nuclear decay
X-ray	0.1 – 100	10^{18}	10^{-16}	core electrons
Ultraviolet	100 – 400	10^{16}	10^{-18}	valence electrons
Visible	400 – 700	10^{15}	10^{-19}	valence electrons
Infrared	700 – 10^6	10^{13}	10^{-21}	vibrations
Microwave	10^6 – 10^9	10^{10}	10^{-24}	rotations
Radio	10^9 <	<10^{10}	10^{-25}	nuclear spin flips

Table 11-1. Divisions of the electromagnetic spectrum

The light we can see unaided is the visible portion of the electromagnetic spectrum. The range of wavelengths gives rise to what we call color. The combination of all colors in light is known as white light; this is what we get from a light bulb or the sun. The spectrum can be broken into its components using prisms or diffraction gratings. Prisms disperse or break up light into component colors of the spectrum by utilizing light refraction (the bending of light through a material). The bending angle depends on the index of refraction of the material, and the index of refraction is different for different wavelengths.

Diffraction gratings cause light dispersion by a different principle from prisms. The gratings used in this experiment are transparent plastic with very fine lines etched into the surface. This regular array of grooves interacts with the light passing through, causing regions of constructive and destructive interference. Regions of destructive interference are dark. Regions of constructive interference appear as spots of light. The Bragg diffraction equation, shown in Eq. 11-3,

$$n\lambda = d\sin\theta \tag{11-3}$$

relates θ, the angle the light bends, the light wavelength λ, and the spacing of the grooves on the diffraction grating, d. The variable n is the order of the diffraction, which for our experiment is equal to one. Light of different wavelengths have different angles for the appearance of constructive interference, which has the effect of spreading white light into its component wavelengths.

The interaction of light and matter can occur in a variety of ways. Light can be taken in by a substance, in a process called **absorption** which results in an increase in energy of the substance. The substance in an increased energy state (in other words, not the ground state) can release the excess energy in the form of a photon in a process called **emission**. The exact wavelengths of absorption and/or emission can be used to identify, or fingerprint, the substance.

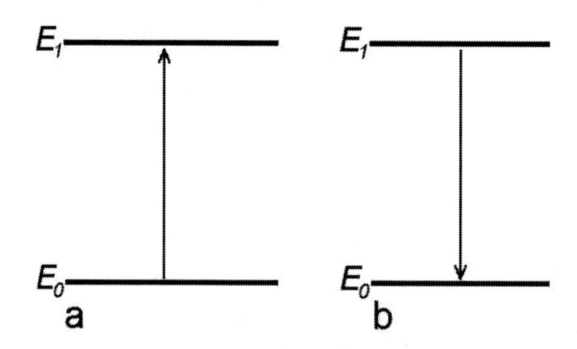

Fig. 11-1. a. Absorption of energy between energy levels of a substance; b. emission of energy between energy levels of a substance.

Experiment 11 Recognizing the "Fingerprints" of Atoms and Molecules

Atomic Emission.
Hydrogen atom has one of the most historically significant spectra since understanding it required inventing quantum mechanics. Unlike the rainbow of white light which comes from the sun, hydrogen only produced select lines. All of the lines in the hydrogen emission spectrum were found to fit the Rydberg equation shown in Eq. 11-4, where λ is the wavelength of the light emitted, R_H is the Rydberg constant, n_2 is the principal quantum number of the higher energy level of hydrogen's electron and n_1 is that of the lower energy level. In inverse nanometers, the Rydberg constant is 0.010972 nm^{-1}.

$$1/\lambda = R_H(1/n_2^2 - 1/n_1^2) \qquad (11\text{-}4)$$

Unfortunately, hydrogen is the only atom with such a "simple" spectrum. Scientists in the early 20th century tried to develop similar formulas for other atoms, but it turned out to be impossible. Other atoms also have line spectra, but the wavelengths cannot be calculated from theory easily.

Molecular absorption spectra. When molecules absorb visible light, the remaining light (transmitted light) is what we see. For example, artificial lemonade drinks are frequently colored with the molecule tartrazine, also known as FD&C Yellow 5. As shown in Table 11-2, yellow light is centered on 570 nm. Tartrazine doesn't absorb much yellow light, though, it absorbs strongly at about 430 nm.

Color	Wavelength(nm)
Red	620-700
Orange	590-620
Yellow	550-590
Green	480-550
Blue	450-480
Indigo	425-450
Violet	400-425

Table 11-2. Wavelengths of the visible color spectrum

While the emission spectra of atoms appear as individual lines of color, the absorption (and emission) spectra of molecules tends to appear as broad peaks. This is partially due to the fact that although the absorptions are primarily due to electrons changing to different orbitals, the electronic change is accompanied by changes in vibrations and rotations of the molecule.

Molecular emission spectra. Four of the more common ways for a molecule to emit visible light are fluorescence, phosphorescence, chemiluminescence, and bioluminescence. An important thing to remember is that a molecule cannot give off light unless it is in an excited state. Hence, there must be an initial excitation of the molecule before it emits.

Fluorescence is a very fast emission of light after the molecule has become excited by light absorption. Typically, molecules will give off the fluorescent light within 1 second, sometimes within nanoseconds of the initial absorption of light. The exciting light is always at shorter wavelength than the emitted light. You may have experienced this when your clothes "glowed" under black lights (which emit UV radiation). Not every molecule is capable of fluorescence; the chemical structures which give off light are referred to as chromophores.

Phosphorescence is generally a much slower emission of light after an excitation by light absorption. This doesn't occur in most molecules. Glow-in-the-dark toys operate by phosphorescence. Most of these contain ZnS, which is a common phosphor. Phosphorescence normally takes longer than fluorescence because in addition to light emission, it also requires a spin flip (*e.g.*, from spin-up to spin-down) of the electron. Initial excitation of the electrons in the ZnS occurs when you "charge" the phosphorescence by exposing it to light. Typically, glow-in-the-dark toys will glow for about 10 minutes, but some newer phosphors last several hours. In 2011, the first fully organic phosphor was reported.[1]

In chemiluminescence and bioluminescence, a fluorescent molecule is excited by a chemical reaction or is created in an excited state by a chemical reaction. In bioluminescence, the chemical reaction occurs in a living organism, such as a firefly, glow worm, or squid. The glow sticks used at Halloween are an example of chemiluminescence. In these light sticks, a fluorescent dye molecule (which varies depending on color) is activated by the reaction of hydrogen peroxide with phenyl oxalate ester. You have to snap the light stick to initiate the glow because the hydrogen peroxide solution and phenyl oxalate ester are initially separated by a breakable container.

Like molecular absorption, fluorescence, phosphorescence, chemiluminescence, and bioluminescence give broad peaks of light rather than the lines of light in atomic emission.

PROCEDURE
Wear your safety goggles.

Part A. The Spectrum of the Sun and of Incandescent Light Bulbs (white light)
a. Obtain a spectroscope.
b. Point the spectroscope (the slit end) at a window with sunlight coming through, or at an incandescent (standard) light bulb. ***Do not point directly at sun.***
c. Peer through the spectroscope and locate the spectrum – these will appear off to the sides of the incoming light source.
d. Rotate the slit at the far end to vertical and rotate the diffraction grating plastic at the near end until you have a rectangular spectrum on the left and right of the slit. (This helps in observing the spectrum. Poor orientations result in a thin line for the spectrum or one which appears in odd locations.)
e. Record observations on the data sheet.

Part B. The Emission Spectrum of the Hydrogen Atom
a. Using the gas emission apparatus, insert the hydrogen gas tube and turn the switch on. Record the color of the light as it appears to your naked eye.
b. Using your spectroscope, determine what colors are emitted from hydrogen. These should appear as lines of color in the same locations those colors appeared from the full spectrum in part A.

Part C. Comparing He and Ne
a. Predict how many lines you expect from He.
b. Insert the He tube in the apparatus and record your observations on the data sheet.
c. Repeat steps a and b for Ne.

Part D. Absorption Spectra (Note 11-1)
a. Obtain cuvettes of yellow and red fruit drink. Estimate their color wavelengths using Table 11-2.
b. Estimate their maximum absorption wavelength using information in the *Introduction*.
c. Using the spectrometer, record the maximum in the transmission spectrum. This should correspond to the color of the solution. How well did you predict the wavelength?
d. Measure the absorbance spectrum using the spectrometer. Record the maximum wavelength(s) of absorbance (where the peak is centered). How well did you predict the wavelength? Is there a single peak or two peaks?

Part E. Chemiluminescence, Fluorescence, and Phosphorescence
    a. Obtain a glowing light stick and a burned out light stick. Record the color of the light stick, and estimate its wavelength.
    b. Using your spectroscope, observe the appearance of the light stick emission.
    c. Using the spectrometer OR the calibrated spectroscope, determine the wavelength of the glow stick's emission.
    d. Place the burned out light stick under a UV light source. What do you observe? How long does the fluorescence remain when the UV source is removed?
    e. Charge a glow-in-the-dark toy under light for several minutes. Determine its glow color and estimate its wavelength.
    f. Examine the toy with the spectroscope and the spectrometer. You will need to be in a very dark room to make observations with your eye directly. What is the appearance of the emission with each? Where is the peak wavelength? What color does this correspond to?

FURTHER READING
1. Bolton, O., Lee, K., Kim, H.-J., Lin, K. Y. and Kim, J. Activating efficient phosphorescence from purely organic materials by crystal design. *Nature Chemistry* **3,** 207–212 (2011).
2. Kuntzleman, T. S., Rohrer, K. and Schultz, E. The Chemistry of Lightsticks: Demonstrations to Illustrate Chemical Processes. *J. Chem. Educ.* **89,** 910–916 (2012).
3. Lisensky, G. C., Patel, M. N. and Reich, M. L. Experiments with Glow-in-the-Dark Toys: Kinetics of Doped ZnS Phosphorescence. *J. Chem. Educ.* **73,** 1048 (1996).
4. O'Hara, P. B., St. Peter, W. and Engelson, C. Turning on the Light: Lessons from Luminescence. *J. Chem. Educ.* **82,** 49 (2005).
5. F. B. Bramwell, S. Goodman, E. A. Chandross and M. L. Kaplan; "A Chemiluminescence Demonstration-Oxalyl Chloride Oxidation" *J. Chem. Educ.* **56** 111, (1979)

NOTES
11-1 This portion of the experiment can be performed using a recording spectrometer or a single point spectrometer. If spectrometers are not available, this portion of Experiment 11 may be omitted.

COMMENTS

Experiment 11 Recognizing the "Fingerprints" of Atoms and Molecules

Name _____ Lab Section _____ Date _____

Prelaboratory Assignment: Experiment 11
Recognizing the "Fingerprints" of Atoms and Molecules
Where appropriate, answers should be given to the correct number of significant digits.

1. Calculate the frequency in Hz of yellow light with a wavelength of 589 nm.

2. Calculate the energy of one photon of yellow light with a wavelength of 589 nm.

3. Will blue light have a longer or shorter wavelength than yellow? How do you know?

4. Will blue light have a lower or higher energy than yellow? How do you know?

Experiment 11 Recognizing the "Fingerprints" of Atoms and Molecules

5. Will blue light have a greater or smaller frequency than yellow light?

6. In the Balmer series for hydrogen atom, the n_1 is always 2. Determine the wavelength in nm of the Balmer line with $n_2 = 4$. What type or color of light is this?

Name _____ Lab Section _____ Date _____ 133

REPORT ON EXPERIMENT 11
Recognizing the "Fingerprints" of Atoms and Molecules

Part A. The Spectrum of the Sun and of Incandescent Light Bulbs

Sketch in the space below a diagram of the colors visible in the spectroscope, noting especially where they appear with respect to the end of the spectroscope.

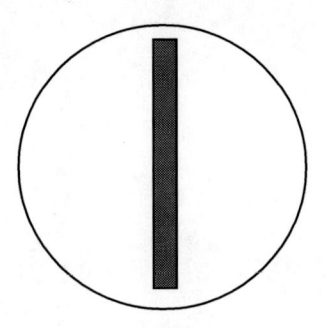

Part B. The Spectrum of the Hydrogen Atom

Color of H emission without spectroscope: _____

Line #	Color	Estimated wavelength (nm)	Balmer transition
1			
2			
3			
4			

Part C. Comparing Ne and He

	helium	neon
Predicted number of lines in spectrum		
Color without spectroscope		
Colors visible with spectroscope		
Number of lines visible		

Experiment 11 Recognizing the "Fingerprints" of Atoms and Molecules

Are the spectra of H, He, and Ne unique, distinct from one another? Explain.

Part D. Absorption Spectra

Fruit drink color	Estimated λ of color	Estimated λ of absorbance	Measured λ of transmittance	Measured λ of absorbance

Part E. Fluorescence and Phosphorescence

Color of glowing light stick _____ Estimated peak wavelength _____

Appearance of spectrum and measured peak wavelength _____

_____

Appearance of burnt out glowstick under UV light _____

Time until fluorescence disappears when UV removed _____

Color of glow-in-the dark toy _____

Estimated wavelength _____

Appearance of glow-in-the-dark spectrum _____

_____

Experiment 11 Recognizing the "Fingerprints" of Atoms and Molecules

Name _____ Lab Section _____ Date _____   135

QUESTIONS (*Submit your answers on a separate sheet as necessary.*)

1. All the Balmer lines have $n_1 = 2$ in the Rydberg equation. What is the value of n_2 for the red line you observed? Show work.

2. In the Bohr model of the hydrogen atom, the electron orbits the nucleus at different distances, with the farther orbits being higher in energy. Draw on the diagram an arrow representing the Balmer emission with $n_2 = 4$. Justify why you chose your starting and ending points and the direction of your arrow.

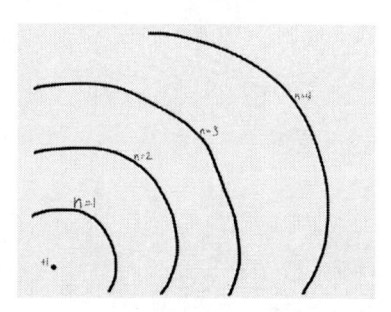

3. A solution of $KMnO_4$ appears purple. You would expect that it is absorbing what color of light?

4. Suggest a reason why we do not use phosphorescent materials as our light bulbs.

Experiment 11 Recognizing the "Fingerprints" of Atoms and Molecules

5. Look up the chemical structure of the chemical lawsone, also known as hennotannic acid, and draw it here. This is the chemical responsible for the color of henna hair and skin dyes. Look up the term "chromophore." Do you see anything in the lawsone structure which may be a chromophore? Explain.

6. Forensic crime scene investigators often use UV "alternative light sources" to get physiological fluids such as saliva to fluoresce. Why doesn't a normal light bulb make the saliva visible?

7. The element helium is known for making balloons float and voices squeaky, but it wasn't discovered until its spectral fingerprint was found in sunlight recorded during the solar eclipse of 1868. Why do you think helium was not easily found on Earth by chemists?

Experiment 11 Recognizing the "Fingerprints" of Atoms and Molecules

Experiment 12
Absorption Spectrophotometry and Beer's Law

OBJECTIVE

To determine the concentration of a copper sulfate solution by absorption spectrophotometry; to analyze a sample of steel for manganese content.

EQUIPMENT

Spectrophotometer, two matched cuvettes or two matched optical cells, two 25-mL burets, assorted small beakers, 150-mL beaker, 250-mL volumetric flask, Bunsen burner, ring stand with 3-in. iron ring.

REAGENTS

50 mL 0.10 M $CuSO_4$, 50 mL stock solution which is 5×10^{-4} M $KMnO_4$ and 2.0×10^{-2} M KIO_4, 1.25 g KIO_4, 1 g of $(NH_4)_2S_2O_8$, 10 mL concentrated HNO_3, 10 mL of concentrated H_3PO_4.

SAFETY AND DISPOSAL

Refer to the MSDS information available online when working with $CuSO_4$,[1] $KMnO_4$,[2] KIO_4,[3] $(NH_4)_2S_2O_8$,[4] HNO_3,[5] and H_3PO_4.[6] Disposal for all compounds must be in accordance with local, state and federal regulations. Disposal for acid solutions should be into a labeled laboratory waste container for acids. Disposal for inorganic salts should be into a labeled laboratory waste container for solid inorganic waste.

INTRODUCTION

If a colored substance is dissolved in a suitable solvent, the intensity of color of the resulting solution will depend on the concentration of the dissolved substance. This fact is the basis for a method of chemical analysis known as colorimetry. The phenomenon of color is the result of absorption and/or reflection of light of specific wavelengths. For example, white light, such as ordinary daylight, consists of electromagnetic radiation having wavelengths ranging from 360 nm (nanometers) to 700 nm. When white light strikes an object, radiation of certain wavelengths is absorbed. The radiation that is reflected from or transmitted through the object will not contain all of the wavelengths originally present in the incident white light, and the object will appear to have color. An object that absorbs radiation in the red region of the visible spectrum will appear green because the light leaving it is rich in the blue and yellow wavelengths. Although the term *color* customarily refers to visible light, many substances absorb radiation in other regions of the electromagnetic spectrum and can be thought of as also having "color." The study of the various electronic and molecular processes responsible for the absorption of electromagnetic radiation is called absorption spectroscopy. These processes require specific amounts of energy; hence, radiation absorbed will have correspondingly specific wavelengths.

Instruments used to measure the intensity of color transmitted or absorbed by a substance are called colorimeters. A spectrophotometer (or spectrometer) is a colorimeter which can continuously vary the wavelength of the light source so that the sample can be observed at selected wavelengths. The construction and operation of a typical spectrometer are described in Appendix C. In this experiment, a spectrophotometer will isolate a beam of light of a particular wavelength, pass it through a sample, and measure the intensity of the light transmitted by the sample, I_t, relative to the intensity of the light incident upon the sample, I_o.

The amount of light of a specific wavelength absorbed by a sample depends on the concentration of absorbing substance in the sample, the thickness of the sample, and the chemical characteristics of the absorbing species. When the sample is in solution, the relationship between these factors is expressed in terms of Beer's law:

$$\log (I_t/I_o) = -\varepsilon bc \tag{12-1}$$

where I_t/I_o is the fraction of the incident light which is transmitted, ε is a constant characteristic of the absorbing species, b is the thickness of the solution, and c is the concentration. When the concentration is expressed in moles per liter and the solution thickness is in centimeters, ε is called the molar extinction coefficient of the absorbing substance.

The absorbance, A, of a sample is defined by the expression

$$A = -\log(I_t/I_o) = \log (I_o/I_t) \tag{12-2}$$

Hence,

$$A = \varepsilon bc. \tag{12-3}$$

For a solution of a given thickness containing a colored substance, this equation for absorbance has the form of an equation for a straight line with zero intercept, y = mx. Therefore, a plot of absorbance versus concentration should yield a straight line having a slope = εb. Experimental data for such plots are obtained by measuring the absorbance of a series of solutions of known concentration. Then the concentration of an unknown solution of the same substance can be found by measuring its absorbance and reading the corresponding concentration of the straight-line plot. Plots of this type, based on Beer's law, are called *working curves* when used in colorimetric analysis.

In practical work, the light transmitted by a solution, I_t, is compared with the light transmitted by a "blank" consisting of the solvent, I_o. Thus, the ratio I_t/I_o can be attributed solely to the absorption of light by the colored substance in the solution.

For a specified concentration and thickness, the absorbance of a substance depends on the wavelength of the incident light. If the absorbance of a solution is measured over a range of wavelengths and these values are plotted against the wavelength, the resultant graph is called the *absorption spectrum* of the solution. Such a plot is shown in Fig. 12-1. Portions of the curve where the absorbance rises to maximum values are called absorption bands. In Fig. 12-1, an absorption band is located at 525 nm. Every colored substance has at least one absorption band between 360 and 700 nm. There may be other bands in the ultraviolet and in the infrared regions of the spectrum as well. For best results, a colorimetric analysis should be made using light with a wavelength corresponding to the center of an absorption band. The application of Beer's law has certain limitations. In many cases, the law holds exactly over a limited range of concentrations. It is most accurate for dilute solutions of a single absorbing species. If the sample contains more than one substance which absorbs at or near the wavelength of the incident light beam, the true concentration will not be that given by a simple working curve based on Beer's law. Also, deviations from Beer's law may occur when the absorbing substance is

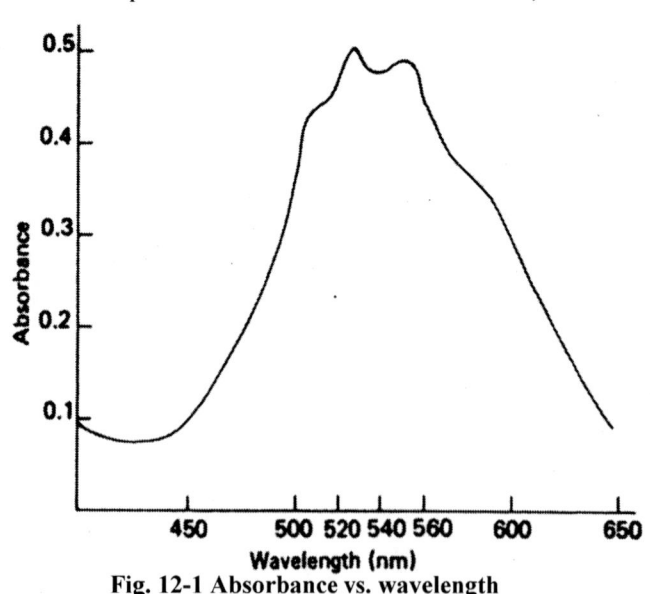

Fig. 12-1 Absorbance vs. wavelength

Experiment 12 Absorption Spectrophotometry

involved in an equilibrium reaction in solution. When such a solution is diluted, the equilibrium may shift so that the change in concentration cannot be predicted in a straightforward manner. Working curves for such a system may not be linear and may have a nonzero intercept. On the other hand, under conditions where Beer's law applies, spectrophotometric analysis is relatively simple and nondestructive. A much smaller sample is required than is necessary for the usual methods of gravimetric analysis or volumetric analysis.

In the first part of this experiment, a spectrophotometer will be used to find the absorption spectrum of an aqueous solution of copper sulfate. After a working curve has been prepared by measuring the absorbance of $CuSO_4$ solutions of known concentration, the concentration of a solution of copper sulfate of unknown concentration will be determined. In the second part of the experiment, the percent of manganese in a sample of steel will be determined by oxidizing the manganese present to permanganate, MnO_4^- and adjusting the resulting solution to a known volume. The concentration of MnO_4^- in the solution will then be found spectrophotometrically by comparison with a working curve for permanganate solutions.

To prepare the steel sample for analysis you must dissolve it in nitric acid. A mild oxidizing agent such as ammonium persulfate, $(NH_4)_2S_2O_8$, is added to oxidize any carbon compounds present, because they might absorb in the same wavelength region as the MnO_4^- ion. The manganese present is then converted to permanganate by means of potassium meta-periodate, a very strong oxidizing agent. Phosphoric acid is added to the final solution in order to convert the Fe^{3+} ions to a colorless complex, $Fe(PO_4)_2^{3-}$.

PROCEDURE

Wear your safety goggles throughout the experiment. Read the instructions for understanding a spectrophotometer in Appendix C. Get detailed instructions on the operation of your spectrophotometer.

Determination of the Copper Ion Concentration

Obtain about 50 mL of 0.10 M $CuSO_4$ stock solution in a clean, dry beaker. Fill a cuvette with this solution and fill another cuvette with distilled water. Insert them into the sample compartment of the spectrophotometer and measure the absorbance of the solution at 30-nm intervals from 400 nm to 700 nm. In the region where the absorbance is highest, take additional readings at 10-nm intervals. Plot a graph of absorbance versus wavelength and attach it to the report on this experiment. Record the wavelength of maximum absorbance on your data sheet. To prepare a working curve, or Beer's law plot, set the wavelength selector of the spectrophotometer at the wavelength where maximum absorbance was observed for the 0.10 M $CuSO_4$ solution. Prepare six solutions of known Cu^{2+} concentrations in the following manner. Using two burets, dilute in turn 2, 4, 6, 8, 12, and 16 mL of the 0.10 M $CuSO_4$ solution with enough distilled water to make 20-mL volumes of each solution. Calculate and record the molarities of the solutions. Measure and record the absorbance of each solution at the selected wavelength. *Note,* when filling the sample cuvette with a new solution, rinse it first with distilled water and then rinse it three times with 1-mL portions of the solution to be measured.

Plot a graph of the absorbance of the solutions versus their concentrations. This plot, which should be a straight line, is the working curve to be used to determine Cu^{2+} concentrations in aqueous copper sulfate solutions of unknown concentration. Attach the plot to the report on this experiment.

Obtain a copper sulfate solution of unknown concentration. Record its code number on your data sheet. Measure its absorbance and determine the concentration of Cu^{2+} in the solution by comparison with the working curve.

Determination of the Percent of Manganese in Steel

Obtain 50 mL of a 5.0×10^{-4} M stock solution of potassium permanganate containing potassium periodate. Using two burets, prepare six solutions of known MnO_4^- concentration by diluting in turn 2,

4, 6, 8, 12, and 16 mL of the stock solution with enough distilled water to make 20-mL volumes of each solution. Calculate and record the molarities of each solution.

Dissolve 0.25 g of KIO_4 in 50 mL of distilled water for use as a solvent blank. Measure the absorbance at 525 nm of each of the MnO_4^- solutions relative to the blank and record the data on your report sheet. (Whenever the sample cuvette is filled with a new solution, it must be rinsed once with distilled water and then three times with 1-mL portions of the solution to be measured.) Plot a working curve for MnO_4^- solutions and attach it to the report on this experiment.

Secure a sample of steel and record its identification number on your data sheet. Weigh out 0.500 ± 0.002 g of the sample on an analytical balance. Transfer the weighed sample to a 150-mL beaker set on a wire gauze upon an iron ring above a Bunsen burner under the hood. Add 40 mL of distilled water to the sample and then *carefully* add 10 mL of 15 M HNO_3. After the initial reaction subsides, boil the mixture gently for 2 or 3 minutes in order to thoroughly extract all manganese from any insoluble residue. Allow the solution to cool below boiling and add 1 g of $(NH_4)_2S_2O_8$ slowly in small amounts so that the solution does not froth or boil over the sides of the beaker. At this point, some of the manganese may precipitate as MnO_2. However, continue adding the $(NH_4)_2S_2O_8$. Dilute the mixture to approximately 100 mL with distilled water and add 10 mL of concentrated H_3PO_4 followed by 0.4 to 0.5 g KIO_4.

Boil the mixture for several minutes until all of the KIO_4 crystals have dissolved. Allow the mixture to cool and add another 0.4 to 0.5 g KIO_4 and boil for 2 minutes more, (Note 12-1).

When the mixture has cooled, transfer it *quantitatively* to a 250-mL volumetric flask. (Review the material on transferring liquid samples in the Laboratory Equipment and Techniques section.) Adjust the final volume to exactly 250 mL with distilled water.

Fill a cuvette with some of the MnO_4^- solution prepared from the steel sample. Fill a second cuvette with the solution used for a blank in the preparation of the working curve. Measure the absorbance of the unknown solution and determine the concentration MnO_4^- from the working curve. Calculate the mass of manganese present in 250 mL of the solution. Calculate the percent by mass of manganese in the steel sample.

Put all solutions used in this experiment into the waste containers specifically set up to receive them.

NOTES

12-1. Any residue of KIO_4 will dissolve completely upon dilution.

FURTHER READING

1. https://fscimage.fishersci.com/msds/05690.htm ($CuSO_4$, last accessed July, 2019)
2. https://fscimage.fishersci.com/msds/19520.htm ($KMnO_4$, last accessed July, 2019)
3. https://www.fishersci.com/msds?productName=AC197760050 (KIO_4, last accessed July, 2019)
4. https://msds.orica.com/pdf/shess-en-cds-010-000031026301.pdf (($NH_4)_2S_2O_8$, last accessed July, 2019)
5. http://www.labchem.com/tools/msds/msds/LC17840.pdf (HNO_3, last accessed July, 2019)
6. http://www.labchem.com/tools/msds/msds/LC18640.pdf (H_3PO_4, last accessed July, 2019)

Name _____ Lab Section _____ Date _____

Prelaboratory Assignment: Experiment 12
Absorption Spectrophotometry and Beer's Law

1. What do the following terms stand for in Beer's Law:
   a. I_t

   b. I_o

   c. b

   d. c

   e. ε

2. What safety precautions must be observed when working with solutions of $CuSO_4$ and with solutions of $KMnO_4$?

3. Why is it necessary to "rinse once with distilled water than three times with 1-mL portions of the solution to be measured" whenever a sample cuvette is filled with a new solution?

Experiment 12 Absorption Spectrophotometry

4. Indicate whether each of the following experimental "blunders" would produce a high result, a low result, or make no difference in the value obtained for the concentration of $CuSO_4$ unknown solution.

   a. The wavelength of the observed maximum is recorded as a value less than the observed wavelength.

   b. In making the "working curve" the $CuSO_4$ solution concentrations are twice as large as reported.

   c. The unknown solution is spilled in placing it in the cuvette. Additional unknown solution is added to the cuvette.

Experiment 12 Absorption Spectrophotometry

Name _____ Lab Section _____ Date _____   143

REPORT ON EXPERIMENT 12
Absorption Spectrophotometry and Beer's Law

DATA AND RESULTS
Determination of the Copper Ion Concentration

DATA FOR ABSORPTION SPECTRUM OF CuSO$_4$ SOLUTION

Wavelength (nm)	Absorbance	Wavelength (nm)	Absorbance

wavelength of maximum absorption (from graph) _____

DATA FOR WORKING CURVE OF CuSO$_4$ SOLUTION

Concentration (M)	Absorbance	Concentration (M)	Absorbance

Unknown code number   _____

Absorbance of unknown   _____

Concentration of Cu^{2+} in unknown   _____

Experiment 12 Absorption Spectrophotometry

DATA FOR WORKING CURVE OF MnO$_4^-$ SOLUTIONS

Concentration (M)	Absorbance at 525 nm	Concentration (M)	Absorbance at 525 nm

Determination of the Percent of Manganese in Steel

unknown number _____

mass of sample _____

absorbance of permanganate solution from
unknown steel sample _____

concentration of MnO$_4^-$ (from working curve) _____

moles of MnO$_4^-$ in 250 mL of solution _____

mass of manganese in 250 mL of solution _____

percent by mass of manganese in steel sample _____

Show the details of all calculations; use extra sheets if necessary.

QUESTIONS
(Submit your answers on a separate sheet as necessary.)

1. Calculate the molar extinction coefficient, ε, for copper ion at the wavelength of maximum absorption by assuming an effective solution thickness of 1.0 cm.
2. Is it possible to calculate the molar extinction coefficient from a single absorbance measurement on one solution of known concentration? What is the advantage of determining this constant from the slope of a graph of absorbance versus concentration?
3. If a solution obeys Beer's law, by what factor does the absorbance change when a 1-cm cuvette is replaced by a 2-cm cuvette?
4. If a solution obeys Beer's law, by what factor does the ratio I_t/I_o change when a 1-cm cuvette is replaced by a 2-cm cuvette?
5. Write balanced equations for all of the chemical reactions involved in the preparation of a steel sample for analysis for manganese. Assume that any carbon present in the sample was in the form of elemental carbon.

Experiment 13
Solution Concentration

OBJECTIVES
To determine the concentration (molarity) of a solution by evaporation of the solvent; to express the calculated concentration in units of molality, mole fraction and mass percent.

EQUIPMENT
Three 125-mL Erlenmeyer flasks, Bunsen burner or hot plate, 5-mL pipette, analytical balance.

REAGENTS
Your lab instructor will issue previously prepared samples of NaCl or KCl with "unknown" concentrations.

SAFETY AND DISPOSAL
Refer to MSDS sheets online when working with KCl[1] or NaCl.[2] Disposal for all compounds must be in accordance with local, state and federal regulations. Solutions of KCl and NaCl may be safely washed down the sink with excess water.

INTRODUCTION
Solutions are homogeneous mixtures of two or more substances. Most commonly there are two components: **solute** and **solvent.** Solutions often serve as a means for obtaining and using the solute in a chemical system. Thus it is frequently necessary to know the exact amount of solute present in the solution or with a given amount of solvent. The concentration relates the amount of solute (as grams or moles) to the amount of solution or solvent (as volume or mass). It is important to remember that the concentration of a solution is a function of both the solute and the amount of solution or solvent. For example, five grams of sugar would make 10 mL of a water solution quite sweet (high concentration) but would hardly be noticed in 1000 mL of water (low concentration). When the amount of solute is expressed in moles and the amount of solution in liters, the concentration term is molarity, (M).

$$\textbf{Molarity} = \frac{\text{moles of solute}}{\text{liters of solution}} \quad (13\text{-}1)$$

In the sugar solutions described above, the same amount of solute (5.0 grams or 0.028 moles) is present in each case, while the amount of solvent differs. In the dilute solution the molarity is 0.028 M (0.028 moles/ 1.0 L) and in the more concentrated solution the molarity is 2.8 M (0.028 moles/ 0.010 L).

Other frequently used concentration terms are mass percent, molality and mole fraction.

$$\textbf{Mass percent} = \frac{\text{mass of solute}}{\text{total solution mass}} \quad (13\text{-}2)$$

$$\textbf{Molality} = \frac{\text{moles of solute}}{\text{kilograms of solvent}} \quad (13\text{-}3)$$

$$\textbf{Mole fraction} = \frac{\text{moles of component}}{\text{total moles all components}} \quad (13\text{-}4)$$

Note the significant difference between **molality** (*m*) and **molarity** (*M*). Molality expresses the concentration as moles of solute per *kilogram of solvent*, while molarity expresses the concentration as moles of solute per *liter of solution*. All four ways of expressing concentration are important but molarity will be used most frequently.

A solution for which the concentration is known exactly is referred to as a **standard solution**. A standard solution can be prepared by dissolving a known amount of a pure substance in a given volume of solution. The concentration of the solution can then be calculated using the known amounts of the solute and solution. However, since most solutions must be prepared from substances that are usually impure, the exact concentration of the solution must be determined experimentally. This process of **standardization**, usually involves the technique of titration. In a titration, one solution is slowly added to a known amount of a second solution until a particular chemical equivalence is achieved. The point of chemical equivalence (endpoint) is noted by a particular event such as a color change or the formation of a precipitate. The specific change observed depends on the chemical reaction. For many systems a small quantity of another chemical (an **indicator**) must be added to produce a clearly visible change.

In this experiment, another method will be used to determine the concentration of a solution. A known volume of a solution will be weighed and the solvent (water in this experiment) will be evaporated. Since the solute in this particular experiment is not volatile, the mass of the residue will be that of the solute. With this information, the molarity, molality, mass percent and mole fraction of the solute in the solution can be calculated.

Procedure:
Wear your safety goggles at all times in the laboratory.

Clean three 125-mL Erlenmeyer flasks. Dry the flasks thoroughly by gentle heating with a Bunsen burner. Obtain an unknown solution and a 5-mL pipette. When cool, weigh each flask carefully to ± 0.01 g.

Rinse the pipet with 2 small portions of the unknown solution. Discard these washings in the appropriate waste container as directed by the instructor. Into one of the flasks, carefully pipette a sample of 5 mL of the unknown. Repeat this step by carefully pipetting 10 mL of the unknown into the second flask. Repeat with the third flask by pipetting 15 mL of unknown into the flask. Reweigh the three flasks and calculate the average density of the solution.

Using a Bunsen burner or hot plate, heat the flask containing the 5 mL sample to evaporate the water. Regulate the heating to avoid any excessive splattering when the residue is nearly dry. When the residue is nearly dry, remove the flask from the burner flame and allow the flask to cool.

Repeat the heating process with the flasks containing 10 mL and 15 mL samples. When the flasks are cool, reweigh each one. NOTE: Check for complete drying by reheating each flask for five minutes.

Reweigh the cool flasks. Determine the mass of residue in each flask. Calculate the average mass percent, molarity, molality, and mole fraction of the solute in the unknown solution.

FURTHER READING
1. https://fscimage.fishersci.com/msds/21105.htm (NaCl, last accessed July, 2019)
2. http://www.labchem.com/tools/msds/msds/LC18790.pdf (KCl, last accessed, July, 2019)
3. Penrose, J.F. "A Practical Application of Molality" *J. Chem. Educ.*, **60**, (1983), 63.
4. Schmuckler, J.S. "Solution Concentration" *J. Chem. Educ.*, **59**, (1982), 61.
5. Toby, S. "A Molality-Molarity Paradox?" *J. Chem. Educ.*, **36**, (1959), 230.

Name _____ Lab Section _____ Date _____ 147

Prelaboratory Assignment: Experiment 13
Solution Concentration

Where appropriate, answers should be given to the correct number of significant digits.

1. What is meant by the following terms:

    a. Molality

    b. Mass percent

    c. Mole Fraction

    d. Molarity

2. If the density of a 700.0 gram H_2SO_4 solution is 1.84 g/cm, what are its molarity and molality?

3. Indicate whether each of the following experimental mistakes would produce a high result, a low result, or make no difference in the value obtained for the concentration of $CuSO_4$ unknown solution.

Experiment 13 Solution Concentration

a. In the first flask, the contents are not completely dried before weighing.

b. In transferring the unknown concentration to the flask, some material is spilled onto the benchtop.

c. The flasks are weighed while they are still cold.

Name _____ Lab Section _____ Date _____   149

REPORT ON EXPERIMENT 13
Solution Concentration
Where appropriate, answers should be given to the correct number of significant digits.

Data Sheet
Unknown solution # _____   Solute: _____

	Trial 1	Trial 2	Trial 3
mass of flask	_____ g	_____ g	_____ g
mass of flask and unknown solution	_____ g	_____ g	_____ g
volume of solution added	_____ mL	_____ mL	_____ mL
mass of flask and nonvolatile residue first heating	_____ g	_____ g	_____ g
mass of flask and nonvolatile residue second heating	_____ g	_____ g	_____ g

Density Calculations

	Trial 1	Trial 2	Trial 3
mass of solution	_____ g	_____ g	_____ g
volume of solution	_____ mL	_____ mL	_____ mL
density of solution	_____ g/mL	_____ g/mL	_____ g/mL

Average density _____ g/mL

Mass Percent Calculations

	Trial 1	Trial 2	Trial 3
mass of solution	_____ g	_____ g	_____ g
mass of solute	_____ g	_____ g	_____ g
mass percent of solute	_____ %	_____ %	_____ %

Average mass percent _____ %

Experiment 13 Solution Concentration

Molarity Calculations

	Trial 1	Trial 2	Trial 3
Mass of solute	_____ g	_____ g	_____ g
Moles of solute	_____ mol	_____ mol	_____ mol
Volume of solution	_____ mL	_____ mL	_____ mL
Molarity (M)	_____ M	_____ M	_____ M

Average molarity _____ M

Molality Calculations

	Trial 1	Trial 2	Trial 3
Moles of solute	_____ mol	_____ mol	_____ mol
Mass of solvent	_____ g	_____ g	_____ g
Molality (m)	_____ m	_____ m	_____ m

Average molality _____ m

Mole Fraction Calculations

	Trial 1	Trial 2	Trial 3
Moles of solute	_____ mol	_____ mol	_____ mol
Moles of solvent	_____ mol	_____ mol	_____ mol
Mole fraction of solute (x)	_____ x	_____ x	_____ x

Average mole fraction _____

QUESTIONS *(Submit your answers on a separate sheet as necessary.)*

1. How will the calculated molarity of the solution be affected if solid particles splatter out of the flask during the drying process? Explain.

2. How will the calculated molarity be affected if the solute residue is not dried thoroughly? Explain.

3. Give clear directions for the preparation of the following aqueous solutions:
    a) one liter of a 5.0% solution of NaCl.
    b) 10 liters of a 6.0 M NaOH solution.

4. A sulfuric acid solution has a density of 1.8 g/mL and is 87% H_2SO_4 by mass. What are the molarity and the molality for this solution?

Experiment 13 Solution Concentration

Experiment 14
Molecular Geometry

OBJECTIVES
To correlate Lewis dot structures with corresponding VSEPR geometries; to visualize the VSEPR geometries. Recognize the VSEPR structure within an extended chain system.

EQUIPMENT
Molecular model kits containing at least the following:

tetrahedral; (carbon)	3	trigonal bipyramidal	1
angular; (oxygen)	1	octahedral	1
(hydrogen, etc.)	6	sticks (for single bonds)	8
(chlorine, fluorine)	2	springs (for double/triple bonds)	3

INTRODUCTION
I. Lewis Structures.
Drawing a Lewis Structure is a very orderly procedure. Trial and error occasionally works, but takes longer.

A. Standard Lewis Structures.

1. Count the valence electrons. For example, in the molecule CF_4, since carbon is from Group 14(4A), it has 4 valence electrons. Fluorine from group 17(7A) has 7 valence electrons. The total number of valence electrons is $4 + 4(7) = 32$. For ions, a negative charge indicates extra electrons and a positive charge indicates fewer electrons.

2. Draw the line structure, with the central atom in the middle. The central atom is *usually* the first atom, unless hydrogen is listed first. Then it would be the next atom afterwards. In CF_4, the central atom is the carbon. The skeletal structure has a carbon drawn with the four fluorine atoms arranged around it.

3. Draw bonds between the atoms (Fig. 14-1). Each bond is an electron pair (2 electrons which are shared). Subtract the bonding electrons from the total to see how many remain. $32 - 4(2) = 24$.

4. Use the remaining valence electrons to complete unfilled octets, starting with the outside atoms (Fig. 14-2). Remember that H (and He) does NOT get an octet; H can only have 2 electrons. Recount to be sure you haven't added extra electrons or left any out.

5. If you don't have enough electrons to fill all the octets, you will have to make double or even triple bonds.

Fig. 14-1. Line structure with bonds added.

Fig. 14-2. Structure including bonds and lone pairs.

B. Special cases

1. *Resonance.* This special case arises when you have a CHOICE about where to put a double bond. As an example, consider the molecule SO_2. There are 18 valence electrons for the molecule. The skeletal structure has the central atom sulfur in between the two oxygen atoms. Four electrons are used to draw the bonds between S and the O's. This leaves 14 electrons to distribute as lone pairs, beginning with the outer oxygen atoms (see Fig. 14-3).

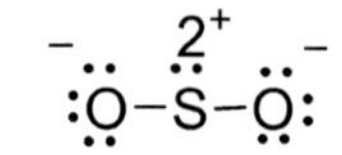

Fig. 14-3. Structure for SO_2 including bonds and lone pairs. Sulfur still lacks an octet. Note the formal charges on the atoms.

2. With all the valence electrons used up, the sulfur still does not have an octet of electrons. The structure in Fig. 14-4 although neutral is not the best structure and therefore must be resolved. The problem is solved using resonance to delocalize electrons and give the best possible structure. A double bond must be created from one of the oxygen lone pairs. But is it created from the left oxygen or the right oxygen? In fact, both answers are correct. Resonance is indicated using curved arrows to show the movement of electrons between atoms.

   The final structure is considered to resonate between the two possible choices. Resonance structures are shown with a double-headed arrow between them, indicating the true structure is a hybrid of the possible choices (see Fig. 14-4).

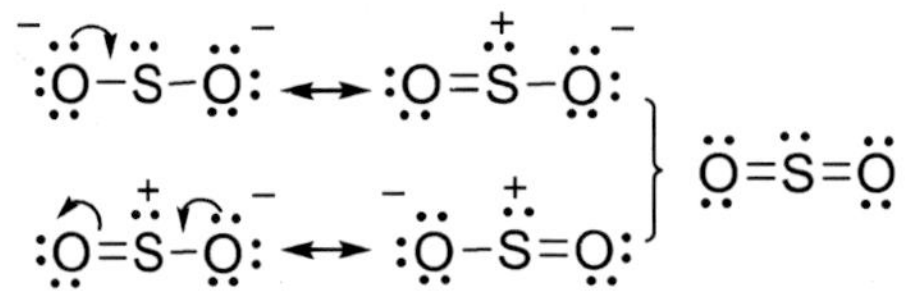

Fig. 14-4. Resonance structures for SO_2. Sulfur has a full octet. The best possible structure is the one in which there is a double bond between sulfur and both oxygen atoms.

3. *Radicals:* Some molecules have an odd number of electrons, indicating that satisfying all octets is impossible. Typically, these are molecules with odd numbers of nitrogen atoms; one example is NO, which has 11 valence electrons. In this case, the nitrogen (less electronegative than oxygen) will only have 7 valence electrons. Species with one unpaired electron are called radicals and tend to be more reactive than standard molecules which have all electrons paired.

   Occasionally, a molecule will have two unpaired electrons. It is not possible to convey this phenomenon with standard Lewis structures. The most common biradical is the oxygen molecule. The unpaired electrons can be demonstrated experimentally through the magnetic properties of liquid O_2.

4. *Expanded Valence Shell:* For third row elements and beyond, in some situations the central atom will have more than 8 electrons bonded to it. This situation is called either expanded octet or expanded valence shell. The rationalization is that atoms with empty d orbitals can use those to have 10 or even 12 electrons in the valence shell.

   For example, the I_3^- ion has $3(7) + 1 = 22$ valence electrons. Four are used to make the bonds between the I atoms, leaving 18 to distribute as lone pairs. (see Fig. 14-5). After completing all octets, there are still 2 valence electrons remaining. These two are put as an extra lone pair on the central I, giving it 10 valence electrons. Xenon is a noble gas, but under extreme laboratory conditions, it will make compounds with fluorine. All these compounds have expanded valence shells for the xenon.

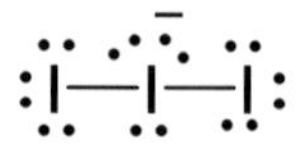

Fig. 14-5. Structure for I_3^- including bonds and lone pairs. The middle atom receives the excess electrons and has a formal charge of -1.

Experiment 14 Molecular Geometry

5. *Incomplete Octets:* A few atoms (B, Be, Al) with fewer than four valence electrons will form compounds with only 4 to 6 valence electrons. BeH_2, BH_3, and AlH_3 all have incomplete octets.

II. VSEPR geometries

The three-dimensional shape of a molecule can be thought of as arising from the repulsion of the electron regions, a model known as valence shell electron pair repulsion (VSEPR). In this model, since all electrons are negative, the electron regions (single bond, double bond, triple bond, lone pair) will separate to the greatest extent possible. In drawing a Lewis structure, the number and types of each electron region are counted. There is a direct relationship between the Lewis structure and the VSEPR geometry of the molecule.

From the Lewis dot structure, count *for each central atom* A the number of:
  a) bonds(X) = single bonds + double bonds + triple bonds = n
  b) lone pairs (E) = m.

The geometry formula can then be expressed as AX_nE_m. For example, on the report sheet, the first two molecules have been completed. For CF_4, there are four single bonds around the central atom C; this leads to the formula AX_4. For the second molecule, there are two central atoms, each of which has four single bonds. Hence, in C_2H_6, the formula is AX_4 for each carbon. In the sulfur dioxide example done previously, in each resonance structure there is a single bond, a double bond, and a lone pair on the sulfur, leading to the formula AX_2E.

The correspondence between the geometry formula and the VSEPR geometry is given in Table 14-1. With only two electron groups on the central atom, the molecule will be linear; the two groups cannot get any farther apart than 180°. When there is a total of four electron groups (combination of bonds X and lone pairs E), the underlying shape is tetrahedral. Since lone pairs are invisible, when the geometry formula is AX3E, the shape is tetrahedral with one bond missing. The result is called trigonal pyramidal. The bent shape arises when two bonds are "invisible."

Electron groups	Geometry formula	VSEPR geometry	angle between groups	hybridization	polarity
2	AX_2	linear	180	sp	nonpolar
3	AX_3	trigonal planar	120	sp^2	nonpolar
3	AX_2E	bent	120	sp^2	polar
4	AX_4	tetrahedral	109	sp^3	nonpolar
4	AX_3E	trigonal pyramidal	109	sp^3	polar
4	AX_2E_2	bent	109	sp^3	polar
5	AX_5	trigonal bipyramidal	120, 90	sp^3d	nonpolar
6	AX_6	octahedral	90	sp^3d^2	nonpolar

Table 14-1. Correspondence among geometry formula, VSEPR geometry, hybridization and polarity

III. Hybridization

In the Valence Bond Model, the valence orbitals of an atom are used to make its bonds. For example, carbon, in row 2, has a $2s$ and three $2p$ orbitals. In addition, carbon has four valence electrons. The four valence orbitals can be combined together to make sp^3 hybrid orbitals, each with one electron for pairing with electrons from another atom. The correspondence between electron groups and hybridization is shown in Table 14-1.

In a molecule with a double bond such as SO_2, there are only three electron groups around the central sulfur, with the geometry formula AX_2E. In the valence bond model, the sulfur has sp^2 hybridization, with each bond and the lone pair being formed from a hybrid orbital. The remaining orbital is not hybridized, but instead is used to form the double bond to an oxygen atom. Expanded

Experiment 14 Molecular Geometry

valence shell systems require *d* orbitals as part of the hybridization. Five electron groups can be formed from sp^{3}d hybrids. Six bonds require sp^{3}d^2 hybridization.

IV. Bond polarity and molecular polarity

When two atoms of the same element are covalently bonded to each other, the shared pair of electrons is shared evenly. When two different elements are covalently bonded, the difference in electronegativity results in the more electronegative atom receiving a larger percentage of the electron pair. This results in a polar bond, which has a partial negative charge on the more electronegative atom and a partial positive charge on the other atom. For example, in HCl, the chlorine is more electronegative and gets a partial negative charge, (see Fig. 14-6). The bond dipole is typically pictured as a vector with one end appearing as a + and the negative end appearing as an arrowhead.

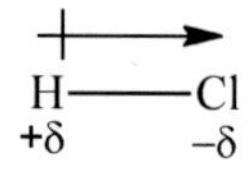

Fig. 14-6. HCl bond polarity

When a molecule has several polar bonds (as in CF$_4$), the molecule will be polar if the bonds are arranged unsymmetrically but will be nonpolar if the molecule is symmetric such that each bond dipole is cancelled out by the others. The geometries which lead to polar and nonpolar molecules are shown in Table 14-1.

Another factor leading to molecular polarity is if the atoms on the central atom are different. For example, CH$_2$Cl$_2$ has four bonds around the carbon and could be thought of as AX$_4$. Since H and Cl are different atoms, however, it is more realistic to think of it as AX$_2$Y$_2$, with a polarity similar to that of AX$_2$E$_2$.

V. Extended systems

Many molecules have a series of central atoms connected together. For example, the molecule C$_3$H$_8$, or propane, has a structural formula CH$_3$CH$_2$CH$_3$. The structural formula clarifies the fact that the molecule is a chain of three carbons, each surrounded by hydrogens. The Lewis structure is as shown in Fig. 14-7.

Each carbon has the geometry formula AX$_4$; the bonds between carbons are counted for both carbons. Each carbon is tetrahedral, with sp^3 hybridization. Hence, in extended systems, each central atom's geometry formula, VSEPR geometry, and hybridization must be considered separately.

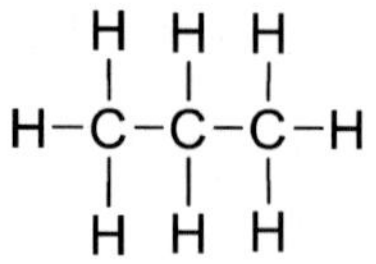

Fig. 14-7. Propane Lewis structure.

PROCEDURE
Part A. Tetrahedral geometries
Build and sketch the tetrahedral molecule CF$_4$. Examine how the appearance of four electron groups changes as more bonds become lone pairs.
Part B. Determining geometries from chemical formulas
Filling in the chart, follow the rules developed in the Introduction part I, write out the Lewis dot structures for the molecules given. Build the molecule using the model kits. Convert the Lewis structure to a geometry formula for the central atom(s), then determine the geometry. Also determine the hybridization and whether the molecule is polar or nonpolar.

Working with a second group, fill in the chart for the larger molecules. Build the larger molecules and sketch their appearances.

FURTHER READING
Ashley S. Jennings, "The VSEPR Challenge: A Student's Perspective," *J. Chem. Educ.* **87**(5), (2010), 462-463. (Includes reference to videos posted by author at http://www.youtube.com/watch?v=i3FCHVlSZc4 (last accessed July, 2019)

Name _____ Lab Section _____ Date _____   155

Prelaboratory Assignment: Experiment 14
Molecular Geometry

1. Using complete sentences, describe the octet rule in your own words. What family of elements satisfies the octet rule without making a compound or molecule?

2. What makes aluminum likely to form an electron deficient compound (with less than a full octet) but the smaller atom fluorine will not?

3. Count the valence electrons for NO_2^- and NH_4^+.

Experiment 14 Molecular Geometry

4. In the geometry formula AX₃E, what do the letters A, X, and E stand for? Be specific.

5. Count the valence electrons in the molecule NCl₃. Which atom is the central atom? How did you decide? Draw its Lewis structure.

6. Count the valence electrons in the molecule SO₂. Which atom is the central atom? How did you decide? Draw its Lewis structure.

Name _____ Lab Section _____ Date _____   157

REPORT ON EXPERIMENT 14
Molecular Geometry

A. Tetrahedral geometries

1. Take a tetrahedral atom (a sphere with 4 holes) and add four single bonds with identical atoms. Which geometry formula and hybridization does this belong to?

   _____

2. Set the molecule on the bench top with one of the lower atoms towards yourself. Sketch the molecule's appearance below.

3. Does it appear that all the bonds are the same? _____

4. Pick up the front atom and move it so it is the one pointing up and a different atom is pointing towards you. Does the molecule look any different from before?

   _____

   Do you think all the bonds are the same in the tetrahedral geometry?

   Explain. _____

   _____

5. Take the protractor and measure the angles between two of the bonds. _____

   Repeat the measurement for any two other bonds. _____

   Do you think the bond angles are the same in the tetrahedral geometry?

   Explain. _____

   _____

   _____

6. Set the molecule on the bench and remove the top atom and bond. Which geometry formula, VSEPR geometry, and hybridization does this correspond to? (Remember, lone pairs are invisible.) _____

   _____

   _____

Experiment 14 Molecular Geometry

Sketch the appearance below.

7. Remove another atom and its bond. To which geometry formula, VSEPR formula, and hybridization does this correspond?

_____

_____

Sketch the appearance below.

Experiment 14 Molecular Geometry

Name _____ Lab Section _____ Date _____   159

B. Determining geometries

9. On the sheets provided, complete the information for the following species.

| CF$_4$ | C$_2$H$_6$ | H$_2$O | SF$_6$ | C$_2$H$_4$ | C$_2$H$_2$ | SO$_3$ |
| CO$_2$ | PO$_4^{3-}$ | NH$_3$ | NH$_4^+$ | HCN | PCl$_5$ | CH$_3$OH |

10. With a second group, make the following molecules: propane: C$_3$H$_8$ (this is a chain) and butane: C$_4$H$_{10}$ (also a chain).

Number of valence electrons	Lewis Structure	Electron Regions		VSEPR Formula	Central Atom Hybridization	Electron Groups and bond angle	Molecular Geometry (sketch with bond angles and name)	Molecular Polarity
		Bonding	Non-bonding					
CF$_4$ 4+4(7)=32	:F–C(F)(F)–F:	4	0	AX$_4$	sp^3	4, 109.5	F–C–F 109.5 (tetrahedral sketch)	nonpolar
C$_2$H$_6$ 2(4) + 6(1) = 14	H–C(H)(H)–C(H)(H)–H	4	0	AX$_4$	sp3	4, 109.5	H–C–C–H Tetrahedral (109) (tetrahedral sketch)	nonpolar
		4	0	AX$_4$	sp3	4, 109.5		

Experiment 14 Molecular Geometry

Number of Valence electrons	Lewis Structure	Electron Regions		VSEPR Formula	Central Atom Hybridization	No. of Electron groups and bond angle	Molecular Geometry (sketch with bond angles and name)	Molecular Polarity
		Bonding	Non-bonding					

Experiment 14 Molecular Geometry

Number of Valence electrons	Lewis Structure	Electron Regions		VSEPR Formula	Central Atom Hybridization	No. of Electron groups and bond angle	Molecular Geometry (sketch with bond angles and name)	Molecular Polarity
		Bonding	Non-bonding					

Experiment 14 Molecular Geometry

Number of Valence electrons	Lewis Structure	Electron Regions		VSEPR Formula	Central Atom Hybridization	No. of Electron groups and bond angle	Molecular Geometry (sketch with bond angles and name)	Molecular Polarity
		Bonding	Non-bonding					

Experiment 14 Molecular Geometry

Name _____ Lab Section _____ Date _____   163

QUESTIONS
(Submit your answers on a separate sheet as necessary.)

1. What is the same about the electron arrangements and geometries for BH_3 and SO_2?

2. Examine Table 14-1. What is the same about all the listed polar geometries?

3. A molecule with the expanded octet geometry formula AX_3E_2 is nonpolar. Explain how this is true.

Experiment 14 Molecular Geometry

4. How many central atoms are in the molecule CH₂Cl-CH₂-CH=NH? What are they?

5. Draw the Lewis structure for the molecule CH₂Cl-CH₂-CH=NH. Determine the geometry for each central atom.

6. The molecule N₂O (in the order NNO) is linear, but NO₂ is bent. Justify this with Lewis structures and the VSEPR geometries.

Experiment 14 Molecular Geometry

Experiment 15
Synthesis of Alum

OBJECTIVES
To prepare an alum from aluminum and to use the colligative property of the melting point to assess the purity of the alum sample.

EQUIPMENT
MelTemp® apparatus, melting point capillaries, hot plate or Bunsen burner, 600-mL beaker, two 100-mL beakers, ring stand, wire gauze, scissors.

REAGENTS
Aluminum, $4M$ KOH, $6M$ H_2SO_4.

SAFETY AND DISPOSAL
Refer to the MSDS information available online when working with H_2SO_4,[1] KOH,[2] and aluminum.[3] Disposal for all compounds must be in accordance with local, state and federal regulations. Rinse all glassware twice with tap water and twice with deionized water. All rinses should be discarded as advised by the instructor.

INTRODUCTION
An **alum** is a hydrated double sulfate salt. The general formula for alum is $M_2(SO_4)_3$ $X_2SO_4 \cdot 24H_2O$, where M is aluminum or a different trivalent metal, and X is a group I metal, or some other monovalent metal. In some cases the ammonium ion can also be X.[4] Table 15-1 gives some alum types and their common uses.

Alum Type	Formula	Chemical Name	Common Use
Chromium Alum	$CrK(SO_4)_2 \cdot 12H_2O$	chromium (III) potassium sulfate dodecahydrate	Used for photograph development
Double salt formula	$Cr_2(SO_4)_3K_2SO_4 \cdot 24H_2O$		
Aluminum Alum	$KAl(SO_4)_2 \cdot 12H_2O$	Potassium aluminum (III) sulfate dodecahydrate	Used for leather tanning and as an astringent
Double salt formula	$Al_2(SO_4)_3K_2SO_4 \cdot 24H_2O$		
Ammonia Alum	$NH_4Al(SO_4)_2 \cdot 12H_2O$	aluminum (III) ammonium sulfate dodecahydrate	Used for fireproofing fabrics
Double salt formula	$Al_2(SO_4)_3(NH_4)_2SO_4 \cdot 24H_2O$		

Table 15-1. Alum Types and Uses

Potassium alum is widely used for household and commercial purposes. Because potassium alum is an astringent (*i.e.* it causes shrinkage), potassium alum is also utilized for **sizing** paper made from wood pulp and for sizing fabrics in the textile industry. The ancients knew of the beneficial properties of alum and it was even used by ancient Romans to purify drinking water.[5]

In this experiment, potassium aluminum (III) sulfate dodecahydrate (the Greek prefix dodeca- means 12) $KAl(SO_4)_2 \cdot 12H_2O$, is prepared by using a common aluminum source. We will use aluminum foil or an aluminum soda can and potassium hydroxide solution. Aluminum metal (foil or can material) rapidly reacts with hot potassium hydroxide solution producing a soluble potassium aluminate salt solution and hydrogen gas (eq. 15-1):

$$2Al(s) + 2K^+(aq) + 2OH^-(aq) + 6H_2O(l) \rightarrow 2K^+(aq) + 2Al(OH)_4^-(aq) + 3H_2(g) \qquad (15\text{-}1)$$

Treating the aluminum tetrahydroxide (aluminate) anion with sulfuric acid causes aluminum hydroxide ($Al(OH)_3$) to precipitate by protonating one of the hydroxide ions on $Al(OH)_4^-$ (eq. 15-2). If this reaction mixture is heated, aluminum hydroxide returns to solution (15-3).

$$2K^+(aq) + 2Al(OH)_4^-(aq) + 2H^+(aq) + SO_4^{-2}(aq) \rightarrow 2Al(OH)_3(s) + 2K^+(aq) + SO_4^{-2}(aq) + 2H_2O(l) \quad (15\text{-}2)$$

$$2Al(OH)_3(s) + 6H^+(aq) + 3SO_4^{-2}(aq) \rightarrow 2Al^{3+}(aq) + 3SO_4^{-2}(aq) + 6H_2O(l) \qquad (15\text{-}3)$$

The following overall net ionic equation. 15-4 is derived from 15-2 and 15-3 and shows the potassium alum product. This product must be cooled in order for it to crystallize from solution because potassium alum is water soluble.

$$K^+(aq) + Al^{3+}(aq) + 2SO_4^{-2}(aq) + 12H_2O(l) \rightarrow KAl(SO_4)_2 \bullet 12H_2O \qquad (15\text{-}4)$$

Once cooling begins, you should be able to detect some crystals being formed. If crystals do not form after cooling, the solution may have to be seeded (*i.e.* a small crystal of pure alum may have to be added to provide a support for the crystals in solution to grow on.)

PROCEDURE
Wear your safety goggles at all times in the laboratory. NOTE: Begin with an exact mass of starting material to synthesize the potassium alum.

Part A. Aluminum Sample Preparation
Cut an approximate 2-inch square of scrap aluminum (soda can or foil). If you are using a soda can, this is a great way to reuse the aluminum that the can is made from. Use caution when cutting the can- pierce it at the end and cut around the hole that you made. Do the same thing to cut the top of the can off. If you unroll the cylindrical piece of aluminum, you will see that is in the shape of a rectangle. Soda cans are typically painted on the outside and coated with plastic on the inside. Clean both sides of the aluminum from the can by scrubbing with steel wool or sand paper. Rinse the aluminum with deionized water. Cut the clean aluminum into small pieces. The reaction rate will be increased if the aluminum pieces are smaller.

Tare a 100-mL beaker. Measure about 0.5 g (+/- 0.01 g) of aluminum pieces into the tared beaker and record the mass.

Part B. Reaction of Aluminum Pieces with KOH
In a fume hood, take the beaker containing the aluminum and add 10-12 mL of 4*M* KOH solution to the aluminum pieces. Swirl the contents of the beaker. KOH can cause serious skin burns, so be careful when adding it and when swirling the beaker.
Bring the beaker from the hood and apply *gentle* heat to the reaction using a Bunsen burner (ring stand and wire gauze will be necessary) or a hot plate. The reaction mixture DOES NOT have to be boiling. You will know that a reaction is taking place because you will be able to see H_2 gas bubbling as the reaction proceeds.

NOTE: The reaction will take approximately 20 minutes. When all of the H_2 gas bubbling has stopped, use tongs to place the beaker on the bench top to begin cooling.

Gravity filter the warm reaction mixture through filter paper into a 100-mL beaker (or 250-mL beaker) to remove the insoluble impurities. If solid particles appear in the filtrate (*i.e.* liquid portion recovered from filtration), repeat the filtration. Rinse the solid left in the gravity funnel with 2-3 mL of deionized water.

Part C. Precipitation of Aluminum Hydroxide

Allow the clear filtrate from step 3 in Part B to cool in the 100-mL beaker. Measure 5 mL of $6M$ H_2SO_4 into 10-mL graduated cylinder. Use a pipette to slowly add in three increments, 2mL, 2mL and 1mL while stirring the reaction mixture. It is important to swirl the mixture in order to ensure proper mixing. **NOTE**: This is a highly exothermic reaction so be very careful. Measure another 5 mL of $6M$ H_2SO_4 into 10-mL graduated cylinder and repeat the process until you have added a total of 10 mL of $6M$ H_2SO_4 to the reaction. Caution: $6M$ H_2SO_4 can cause burns to the skin.

Recall, $Al(OH)_3$ is insoluble. When you see the white $Al(OH)_3$ precipitate in the acidified filtrate, stop adding the $6M$ H_2SO_4. Gently heat (Bunsen burner or hot plate) the mixture until the $Al(OH)_3$ dissolves.

Part D. Crystallizing and Isolating Alum from the Solution

Make an ice bath by halfway filling a large beaker (500- or 600-mL) with ice and water. After everything has dissolved in your 100-mL beaker, remove the solution from the heat. Allow the beaker to cool down for about 2 minutes by placing it on the bench top. *Do not immediately place the beaker into the ice bath as the crystals formed will be very impure.*

After 2 minutes, place the beaker into the ice bath and cool. Alum crystals should begin to form within 20 minutes. If the crystals do not form, scratch the inside of the beaker with a glass rod. The microscopic scratches on the surface of the glass create places for the crystals to begin to grow. You may also need to use a seed crystal (see Introduction above). Your instructor will guide you if your crystals do not form.

Set up a vacuum filtration apparatus and filter the alum crystals from the solution. Wash the crystals with two ice cold, 5 mL portions of a 50:50 ethanol-water solution. It is important to ensure that the ethanol water solution concentration is accurate as alum will dissolve in water but is only moderately soluble in the ethanol-water solution.

Disposal: Rinse all glassware twice with tap water and twice with deionized water. All rinses should be discarded as advised by the instructor.

Part E. Determination of Alum Purity via Melting Point

The melting point of the alum sample can be determined using a melting point apparatus. Your instructor will assist you with the set-up of the apparatus. This should be done when the alum is *totally dry* and may have to be completed in a subsequent lab period. Prior to determining the melting point, you should research the melting point for potassium alum in order to have a reference to compare with your alum's melting point.

Obtain a melting point capillary tube and place finely ground, dry alum about 0.5 cm in the bottom of a melting point capillary tube. To get the alum into the capillary tube, place some alum on a piece of dry filter paper and press the open end of the capillary tube into the alum. Turn the capillary tube and tap it on the bench top to move the alum to the bottom of the tube.

Place the tube containing the sample into the melting –point apparatus. Slowly heat the sample at a rate of about 3°C per minute (if you are using a Mel-Temp® apparatus, the transformer setting should not exceed 2). Carefully watch the alum sample through the viewfinder. The term melting point is a bit deceptive. You are actually measuring a range and not a singular point. The moment the solid begins to melt, record the temperature. The moment the solid finishes melting, record the temperature. Report as the melting range in your data sheet. The closer this range is to the observed range for alum, the purer is your sample.

FURTHER READING
1. http://www.labchem.com/tools/msds/msds/LC25850.pdf (H_2SO_4, last accessed July, 2019)
2. http://www.labchem.com/tools/msds/msds/LC19190.pdf (KOH, last accessed July, 2019)
3. https://www.spectrumchemical.com/MSDS/A3560.pdf (Aluminum, last accessed July, 2019)
4. "Alum" 21 July 2009. HowStuffWorks.com. http://science.howstuffworks.com/alum-info.htm (last accessed July, 2019)
5. Faust, S.D.; Aly, O.M.; *Chemistry of Water Treatment* (2nd ed.), Chelsea, MI: Ann Arbor Press, 1999.
6. "Synthesis and Technique in Inorganic Chemistry (3rd ed.)" Gorilami, G.S.; Raughfuss, T.B.; Angelici, R.J.; University Science Books, 1999.
7. Tesh, K.F. "The Use of Potassium Alum in Demonstrating Amphoterism" *J. Chem. Educ.*, **69**, (1992), 573.
8. Fliedner, L.J. "The Preparation and Preservation of Large Crystals of Chrome Alum" *J. Chem. Educ.*, **9**, (1932), 1453.

Name _____ Lab Section _____ Date _____

Prelaboratory Assignment: Experiment 15
Synthesis of Alum

Where appropriate, answers should be given to the correct number of significant digits.

1. An alum is a double salt consisting of a monovalent cation, a trivalent cation, and two sulfate ions with 12 waters of hydration (water of crystallization) as part of the crystalline structure.

    a. Are the 12 waters of hydration used to calculate the theoretical yield of the alum? Explain.

    b. The 12 waters of hydration are coordinated to the metal ions in the crystalline alum structure. Are the water molecules more strongly coordinated to the monovalent cation or the trivalent cation? Explain.

    c. What might you expect to happen to the alum if it were heated to a temperature lower that the boiling point of water? Explain.

2. Potassium alum, synthesized in this experiment, has the formula $KAl(SO_4)_2 \cdot 12H_2O$; written as a double salt, however, its formula is $K_2SO_4 \cdot Al_2(SO_4)_3 \cdot 24H_2O$. Refer to Table 15.1 and write the formula of sodium alum as a double salt.

3. Answer the following:

    a. What is the technique for fitting a piece of filter paper into a funnel for gravity filtration?

Experiment 15 Synthesis of Alum

b. What is the technique for securing a piece of filter paper into a Buchner funnel for vacuum filtration?

c. Why are the alum crystals washed with an alcohol-water mixture rather than with deionized water?

4. For the synthesis of potassium alum, advantage is taken of the fact that aluminum hydroxide is amphoteric, meaning it can react as an acid (with a base) or a base (with an acid). Complete and balance the following equations showing the amphoteric behavior of aluminum hydroxide

   a. As a base: $Al(OH)_3(s) + H_3O^+ \rightarrow$

   b. As an acid: $Al(OH)_3 + OH^- \rightarrow$

5. An aluminum can is cut into small pieces. A 2.16 g sample of the aluminum chips is used to prepare potassium alum according to the procedure described in this experiment. Calculate the theoretical yield (in grams) of potassium alum that could be obtained in the reaction using the correct number of significant figures. The molar mass of potassium alum is 474.39 g/mol.

Experiment 15 Synthesis of Alum

Name_____ Lab Section_____ Date_____ 171

REPORT ON EXPERIMENT 15
Synthesis of Alum
Where appropriate, answers should be given to the correct number of significant digits.

DATA AND RESULTS

        Mass of aluminum: _____ g

        Mass of alum synthesized: _____ g

        Theoretical yield: _____ g

Percent yield (%): _____

Melting Range

Trial 1	Trial 2
Beginning Temperature: _____ °C	Beginning Temperature: _____ °C
Ending Temperature: _____ °C	Ending Temperature: _____ °C

Average Melting Temperature: _____ °C

What melting point (range) did you find for potassium alum?

How does the observed melting range for the alum that you synthesized compare?

Comment on the purity of your sample based on your experimental melting point.

Experiment 15 Synthesis of Alum

QUESTIONS
(Submit your answers on a separate sheet as necessary.)

Circle the questions that have been assigned.

1. The aluminum sample is not cut into small pieces but rather left as one large piece.
   a. How will this oversight affect the progress of completing the experimental procedure? Explain.

   b. Will the percent yield of the alum be too high, too low, or unaffected by the oversight? Explain.

2. Aluminum pieces inadvertently collect on the filter.
   a. If left on the filter, will the percent yield of the alum be reported as too high or as too low? Explain.

   b. If the aluminum pieces are detected on the filter, what steps would be used to remedy the observation? Explain.

3. In a hurry to complete the synthesis, Bill used $6M$ HCl, also on the reagent shelf, instead of the $6M$ H_2SO_4. As a result, describe what observation would he expect.

4. Andrea used too much sulfuric acid. What observation would she expect? Explain.

5. Explain why the alum crystals are washed free of impurities with ethanol-ice-water mixture rather than with room-temperature deionized water.

6. Explain why the melting point of your prepared alum must either be equal to or be less than the actual melting point of the alum. Consult with your laboratory instructor.

7. Experimentally, how can the moles of the waters of hydration in an alum sample be determined?

8. A greater yield and larger alum crystals may be obtained by allowing the alum solution to cool in a refrigerator overnight or for a few days. Explain.

Experiment 15 Synthesis of Alum

Experiment 16
Kinetics of Bleaching a Food Coloring

OBJECTIVE
  To identify factors affecting the rate of a first order chemical reaction.

EQUIPMENT
  Spectrometer with timing capability or with stopwatch; associated cuvettes; 100-mL graduated cylinder; 1-mL and 10-mL graduated pipettes; 30-mL beaker; computer with graphing program; hot plate; 0 – 100°C thermometer; 400-mL beaker; 18x150-mm test tubes.

REAGENTS
  Part A: Green food coloring (3 drops per student or partner pair). One mL commercial bleach. 90 mL deionized water. *Part B*: Two mL commercial bleach; 100 mL deionized water.

SAFETY AND DISPOSAL
  Refer to the MSDS information available online when working with NaOCl.[1] Disposal for all compounds must be in accordance with local, state and federal regulations. Dispose of any bleach solutions down the drain with copious amounts of water. *Do not combine with waste solutions containing ammonia.*

INTRODUCTION
  Many chemical reactions occur fast enough that we can't detect a time lag. For example, when you add silver nitrate solution to sodium chloride solution, a precipitate forms instantaneously:

$$AgNO_3(aq) + NaCl(aq) \rightarrow AgCl(s) + NaNO_3(aq)$$

This may give the impression that all reactions are fast, but this is not so. How long will it take a car to turn completely to rust? How long does it take to digest your food? The study of the rates of chemical reactions and the factors affecting the rates is known as **chemical kinetics**.

  In this experiment, you will determine factors affecting the rate of a chemical reaction. The reaction you will study is the bleaching of a commercial food coloring dye by household bleach. Household bleach is an aqueous solution of *the oxidizing agent* sodium hypochlorite. As you are probably well aware, bleach will cause colored substances to become colorless. This is because coloring agents are susceptible to oxidation, often through the change of double bonds to single bonds. When oxidized, their structure changes along with their ability to absorb colored light. Commercial food dyes are labeled, for example, FD&C Red No. 2. This indicates the color of the dye and that it has been approved for use with **F**oods, **D**rugs, and **C**osmetics by the Food and Drug Administration (FDA).

  In this experiment, we ask you to determine how various factors affect the rate of bleaching. Factors you should consider are:
- amount of dye
- amount of bleach
- temperature

Experiment 16 Kinetics of Bleaching a Food Coloring

In any experiment, a scientist must be careful to control the variables. Typically, in chemistry, it is recommended that only one variable should be allowed to change at a time to simplify the analysis of how the reaction is affected. We will use the fact that Beer's Law for light absorption is $A = \varepsilon bc$, where A is absorbance of light, ($A=\log(I/I_o)$ where I is the light intensity through the sample and I_o is the light intensity through the blank); ε is the molar absorption coefficient of the substance; b is the path length (1-cm cuvettes will be used) and c is the concentration of the sample in mol/L.

Fig. 16-1 shows changes in time of the absorbance (which is proportional to the concentration, see above) of the food coloring after bleach has been added. Initially, the absorbance is high because there is a lot of food coloring present. As time passes, the absorbance decreases because the colored dye is decreasing in concentration.

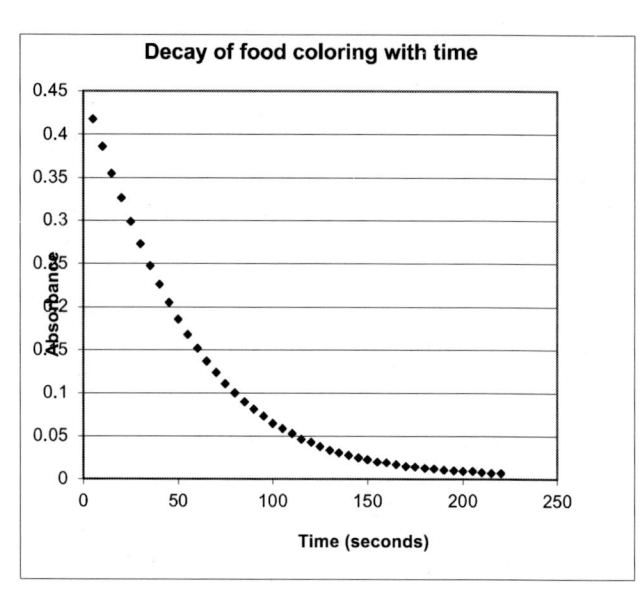

Fig. 16-1. Change in absorbance with time of a reacting food dye.

For a **first order** reaction with the differential rate law:

$$rate = k[X],$$

there are integrated rate laws which are written as:

$$\ln[X]_t = \ln[X]_0 - kt, \quad (16\text{-}1)$$

or

$$[X]_t = [X]_0 e^{-kt}. \quad (16\text{-}2)$$

Examining Eq. 16-1, it can be shown that a plot of $\ln[X]$ vs. time will produce a straight line with a slope of $-k$. A plot of $[X]$ against time should follow Eq. 16-2, and show an exponential behavior.

For a **second order** reaction with a differential rate law of rate= $k[X]^2$, the integrated rate law is:

$$\frac{1}{[X_t]} = \frac{1}{[X_o]} + kt. \quad (16\text{-}3)$$

Eq. 16-3 indicates that for a second order reaction, a plot of $1/[X]$ vs. time will give a straight line with a slope of k.

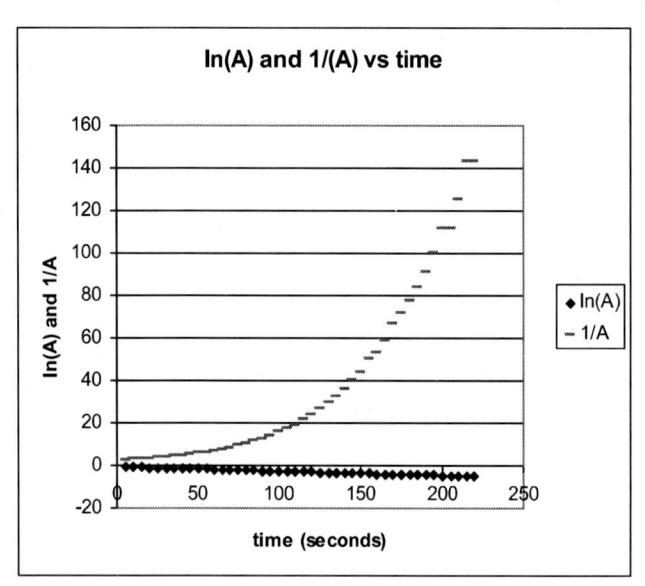

Fig. 16-2. Plots of natural log of concentration (diamonds) and inverse of concentration (dashes) for a reacting food dye with time. As this is a first-order reaction, the natural log shows straight-line behavior and the inverse of concentration is curved.

In Fig. 16-2, the data from Fig. 16-1 has been plotted again to show how ln(Absorbance) and 1/(Absorbance) change with time.

For many reasons, reactions are run with a large excess of one reagent. For example, the reaction

Experiment 16 Kinetics of Bleaching a Food Coloring

$$C + D \rightarrow E$$

may be conducted with C as the limiting reagent and with D in excess. If the reaction is a second order reaction of the type:

$$\text{rate} = k[C][D],$$

the effect of D will not be apparent because its concentration will not change very much during the reaction. Under these conditions, the reaction is **pseudo-first order**, and has the rate law

$$\text{rate} = k'[C],$$

where

$$k' = k[D].$$

The pseudo-first order reaction will behave like a first-order reaction. Changing the concentration of D while letting it remain in excess will change k', but the reaction will still appear to be first order. Decreasing the amount of D until it is nearly stoichiometric will change the behavior to second order in C and D.

PROCEDURE

Wear your safety goggles at all times in the laboratory.

Part A: Observing the characteristics of bleaching.

Make a stock solution of food coloring by adding 3 drops to 90.0 mL of deionized water (dye stock solution). Label and save this solution for use throughout the experiment. Combine 9.0 mL of the dye stock solution with 1.0 mL of bleach in a 30-mL beaker using graduated pipettes. Observe and record the results. You should time the changes (*e.g*, "the solution began to bubble after 45 s"). Determine the time it takes for the dye to completely react. Hypothesize about the effects of changing amounts and temperature.

Part B: Use the spectrometer to quantify the bleaching process.

Follow your instructor's directions to start up your spectrometer, set the wavelength for 610 nm. Use a cuvette of water to "zero" and "blank" the spectrometer. Zeroing a spectrometer calibrates it for no light reaching the detector. Blanking it calibrates the spectrometer for full beam intensity of I_0.

Find the %Transmittance and absorbance of the mixture of 9.0 mL of stock solution with 1.0 mL of deionized water. Configure the spectrometer again according to your instructor's directions to record absorbance every 5 seconds for a total of 250 seconds or until the reaction is complete, depending on your results from Part A.2. Alternatively, prepare to time and record the absorbance by hand each 5 seconds for 50 points. Place 9.0 mL of stock food coloring in a 30-mL beaker. Get 1.0 mL of bleach *ready to add*.

Get your cuvette ready. **At the same time**, add the bleach to the 30-mL beaker and press the "start" button on the computer. QUICKLY transfer your mixture to the cuvette and the cuvette into the spectrometer so the first point (measured at 5 seconds) will be taken. If you miss the first point, the experiment is still valid, but having the first point is preferable.

Save/export the data into a graphing program and create a graph of absorbance vs. time. Your instructor may provide you with additional instructions for this step. Fit the graph to an exponential to determine the rate constant. Your instructor will advise you whether to save a soft or hard copy for your report.

Part C: Changing the amount of food coloring.

What happens to the rate when the amount of food coloring is changed? Choose amounts of dye stock solution, deionized water, and bleach to combine such that the final volume is 10 mL. Keep the amount of bleach constant, but vary the amounts of deionized water and dye stock so that the final volume is 10 mL. Have your instructor approve your setup.

Estimate how long the reaction will take.

Experiment 16 Kinetics of Bleaching a Food Coloring

Repeat the above steps for four different combinations of dye stock, deionized water, and bleach. Remember to add the bleach last, and to start the timer at the same moment you add the bleach to the 30mL beaker.

Part D: Changing the amount of bleach.

What happens to the rate when the amount of bleach is changed? Choose amounts of dye stock solution, deionized water, and bleach to combine such that the final volume is 10 mL. Keep the amount of dye stock constant, but vary the amounts of water and bleach so that the final volume is 10 mL. Have your instructor approve your setup.

Estimate how long the reaction will take. Repeat the above steps for four different combinations of dye stock, deionized water, and bleach. Remember to add the bleach last, and to start the timer at the same moment you add the bleach to the 30-mL beaker.

Part E: Changing temperature.

What happens if the temperature is increased? Prepare a hot-water bath by boiling 200 mL of water in your 400-mL beaker. Remove the beaker from the hot plate onto a safe location such as a wire gauze pad. Place 9.0 mL of dye stock in an 18x150-mm test tube and then place the test tube in the hot-water bath, making sure that its *contents* are well below the surface of the hot-water bath. Measure and record the temperature of the dye stock. Add 1.0 mL of bleach to the dye stock in the test tube and record your observations.

What happens if the temperature is decreased? Prepare a cold-water bath by placing ice up to the 200 mL mark in your 400-mL beaker. Add a mixture of 25 mL of water containing 3 g NaCl. Place the beaker onto a safe location such as a wire gauze pad. Place 9.0 mL of dye stock in a 18 x 150mm test tube and then place the test tube in the cold-water bath, making sure that its *contents* are well below the surface of the cold-water bath. Measure the temperature of the dye stock. Add 1.0 mL of bleach to the dye stock in the test tube and record your results.

FUTHER READING
1. http://www.labchem.com/tools/msds/msds/LC24630.pdf (NaOCl, last accessed July, 2019)
2. Weaver, G. C. & Kimbrough, D. R. Colorful Kinetics. *J. Chem. Educ.* **73**, 256 (1996)
3. Sanger, M. J., Wiley, R. A., Richter, E. W. & Phelps, A. J. "Rate Law Determination of Everyday Processes" *J. Chem. Educ.* **79**, 989 (2002).

Name _____ Lab Section _____ Date _____   177

Prelaboratory Assignment: Experiment 16
Kinetics of Bleaching a Food Coloring

1. What is the main active component of commercial bleaches? Provide the name and the chemical formula.

2. How does a catalyst affect the reaction rate? Does it increase or decrease the rate?

3. For the true first-order reaction A + B → C, if the rate law is: rate = $k[A]$, how will the rate change when the concentration of A is tripled? Explain.

4. The rate law for the reaction A + B → D is rate = $k[A][B]$. If the reaction is second order, how will the rate change when the concentration of B is doubled and the concentration of A is also doubled? Explain.

Experiment 16 Kinetics of Bleaching a Food Coloring

5. Look up the Arrhenius law in your textbook or on the internet and define each term below. What physical or chemical problem does this equation address?

k _____

A _____

E_a _____

R _____

T _____

Name _____ Lab Section _____ Date _____   179

REPORT ON EXPERIMENT 16
Kinetics of Bleaching a Food Coloring

Part A
Record the dye ingredients in the food coloring used for this experiment:

_____

Observations on combining dye stock with bleach:

_____

_____

_____

Estimate of time until reaction is complete: _____

Hypothesis 1: When the concentration of dye stock is decreased, then

_____

_____

because

_____

Hypothesis 2: When the concentration of bleach is decreased, then

_____

_____

because

_____

Hypothesis 3: When the temperature is decreased, then

_____

because _____

Experiment 16 Kinetics of Bleaching a Food Coloring

Part B

9.0 mL dye stock + 1.0 mL of water:

% Transmittance_____   Absorbance_____

Time it took to get reaction cuvette into spectrometer_____

Time for reaction to appear complete_____

Rate constant from graph_____

Part C

_____mL dye stock + _____mL of water + _____mL bleach

Estimate of reaction time_____

Time to get cuvette into spectrometer_____

Time for reaction to appear complete_____

Rate constant from graph_____

Part D

_____mL dye stock + _____mL of water + _____mL bleach

Estimate of reaction time_____

Time to get cuvette into spectrometer_____

Time for reaction to appear complete_____

Rate constant from graph_____

Part E
1. Increased temperature.

How the reaction was performed at increased temperature:

_____

_____

Time for complete reaction: _____

Rate constant from graph: _____

Experiment 16 Kinetics of Bleaching a Food Coloring

Name _____ Lab Section _____ Date _____

2. Decreased temperature.

How the reaction was performed at decreased temperature:
_____

Time for complete reaction: _____

Rate constant from graph: _____

In part A, you wrote down three hypotheses. Compare these to your results below, and discuss any differences.

Hypothesis 1:
_____
_____
_____

Hypothesis 2:
_____
_____
_____

Hypothesis 3:
_____
_____
_____

Experiment 16 Kinetics of Bleaching a Food Coloring

QUESTIONS
(Submit your answers on separate sheets if necessary.)

1.
   a. Make two new plots of the data from part B: one as ln[A] versus time, and the second as 1/[A] versus time.
   b. Which one is a straight-line graph?
   c. From a linear fit or trend line, determine the slope of the straight-line graph and the rate constant.
   d. Is the value the same as you determined in part B?
   e. Do the results show this reaction is first order or second order in the dye concentration?

2.
   a. Compare the time to complete the reaction in parts B and C of the experiment.

   b. Compare the rate constants from parts B and C of the experiment.

   c. Should time of completion and the rate constant change between parts B and C of the experiment? Explain.

3. At what value of positive t do the exponential function e^{-t} and the inverse function $1/t$ become zero? What does this say about "time of completion" for first and second order reactions?

4. The compound which makes carrots orange is β-carotene (see structure below). Explain with reference to its structure how it might be affected by commercial bleach

Experiment 16 Kinetics of Bleaching a Food Coloring

Name _____ Lab Section _____ Date _____   183

5. Cooking speeds up reactions in your food – different ones for meat vs. potatoes. Using outside resources, find the components of each that are affected.
(http://www.smithsonianmag.com/science-nature/why-fire-makes-us-human-72989884/)

6. From a molecular point of view, what can account for the changes in the reaction rate with temperature?

7. Using your data and results from question 1 and your class textbook, determine the half-life of the reaction from part B. How long will it take for a sample of the stock solution to be reduced to one eighth its original value?

Experiment 16 Kinetics of Bleaching a Food Coloring

COMMENTS

Experiment 16 Kinetics of Bleaching a Food Coloring

Experiment 17
pH Measurement and Buffer Solutions

OBJECTIVES
To measure pH by an indicator method and by a pH meter method; to prepare a buffer solution of a prescribed pH; to determine the dissociation constant of a weak acid or a weak base from a pH measurement.

EQUIPMENT
pH meter, six 12x75-mm test tubes, 100-mL graduated cylinder, two 250-mL beakers, medicine dropper, stirring rod, Universal pH paper

REAGENTS
Set of sulfonphthalein indicator dye solutions with color chart; buffer solutions for standardizing pH meter; 1 mL of 0.1 M HCl, 1 mL of 0.1 M NaOH; 0.1 M aqueous solutions of the following: 100 mL each of NH_3, NH_4Cl, $HC_2H_3O_2$, $NaC_2H_3O_2$ and 10 mL each of H_3BO_3, H_3PO_4, NaH_2PO_4, Na_2HPO_4, Na_2CO_3, $NaHCO_3$, Na_2SO_3, $NaHSO_3$, $ZnCl_2$; solutions of appropriate unknowns.

SAFETY AND DISPOSAL
Refer to the MSDS information available online when working with HCl,[1] NaOH,[2] NH_3,[3] NH_4Cl,[4] $HC_2H_3O_2$,[5] $NaC_2H_3O_2$,[6] H_3BO_3,[7] H_3PO_4,[8] NaH_2PO_4,[9] Na_2HPO_4,[10] Na_2CO_3,[11] $NaHCO_3$,[12] Na_2SO_3,[13] $NaHSO_3$,[14] $ZnCl_2$.[15] Disposal for all compounds must be in accordance with local, state and federal regulations. Disposal for these compounds and their derived salts should be into a labeled laboratory waste container for inorganic chemicals. Disposal for unreacted acids should be into a labeled laboratory waste jar for acids. Disposal for unreacted bases should be into a labeled laboratory waste jar for bases.

INTRODUCTION
Many aqueous solutions have an excess of either hydronium (H_3O^+) or hydroxide (OH^-) ions as a result of proton exchange or hydrolysis reactions that occur between the solute and water. Hydronium ion are produced when the solute species donate protons to water molecules, as in the case of the strong acid HCl.

$$HCl + H_2O \rightleftarrows H_3O^+ + Cl^-$$

If the acid is weak, the protonation process occurs only to a limited extent and is expressed as an equilibrium reaction. For example,

$$HNO_2 + H_2O \rightleftarrows H_3O^+ + NO_2^-$$

$$NH_4^+ + H_2O \rightleftarrows H_3O^+ + NH_3$$

The conjugate base of an acid is the residue of the original acid remaining after the proton has been lost. For example, the conjugate bases of HNO_2 and NH_4^+ are NO_2^- and NH_3, respectively. Hydronium ions may also be produced when certain cations hydrolyze in water to form hydro- complexes, as in the following example.

$$Fe^{3+} + 2H_2O \rightleftarrows H_3O^+ + Fe(OH)^{2+}$$

Aqueous solutions of base have an excess of OH⁻ ions. The base may be a hydroxide itself, like the strong base NaOH, or it may generate OH⁻ ions by accepting protons from water molecules as, for example, the weak bases NH_3 and NO_2^-.

$$NH_3 + H_2O \rightleftarrows OH^- + NH_4^+$$

$$NO_2^- + H_2O \rightleftarrows OH^- + HNO_2$$

A protonated base is the conjugate acid of that base. Thus, NH_4^+ and HNO_2 are the respective conjugate acids of the bases NH_3 and NO_2^-.

The pH Scale

A convenient measure for the magnitude of the hydronium ion concentration ($[H_3O^+]$) in an aqueous solution is the pH scale. The pH of a solution is strictly defined as the negative logarithm of the hydronium ion activity. However, in dilute solutions where the activity value approaches that of the molarity, the pH may be approximated by the expression

$$pH \cong -\log[H_3O^+] \tag{17-1}$$

The pH is expressed as a dimensionless quantity that is numerically approximated by the negative logarithm of the number of moles of H_3O^+ present in a liter of solution at equilibrium.
A change of 1 pH unit is equivalent to a tenfold change in the hydronium ion concentration.

pH	$[H_3O^+]$ mole/L
:	:
-1	10^1
0	$10^0 = 1.00$
1	10^{-1}
2	10^{-2}
3	10^{-3}
:	:

In a similar manner, the pOH of a dilute aqueous solution is related to the magnitude of the hydroxide ion molarity as follows.

$$pOH \cong -\log[OH^-] \tag{17-2}$$

A useful relation between the pH and pOH of an aqueous solution at 25°C is called K_w. It may be derived from the value of the ion product for water at this temperature:

$$K_w \cong [H_3O^+][OH^-] \cong 1.00 \times 10^{-14} \text{ at 25°C} \tag{17-3}$$

Taking the negative logarithms of terms in eqn. 17-3 and substituting eqn. 17-1 and eqn. 17-2 yields the simple equation

$$pH + pOH \cong 14.00 \text{ at 25°C} \tag{17-4}$$

Therefore, for pure water at 25°C where $[H_3O^+] = [OH^-] = 1.00 \times 10^{-7}$, the pH = pOH = 7.00. (Distilled water usually has a pH less than 7.00 because of the H_3O^+ ions formed as a result of reactions between water and dissolved CO_2 from the air.)

In this experiment, the indicator method is one of the ways by which you will measure the solution pH. An indicator is generally an organic dye that changes color over a characteristic narrow pH interval. The particular pH range for color sensitivity varies widely for different indicators (see Appendix B, Table

10). Thus, if an appropriate indicator dye is added to a solution, the pH of that solution can be determined to an accuracy of about 0.5 pH unit by observing the resultant color.

Measurement with a pH meter, the second method that will be used, provides an accuracy of about 0.05 pH unit. Although the details of the operational procedure may differ from one instrument to another, the basic principle of operation underlying all pH meters is the same. The solution to be tested is introduced as a component of a special type of galvanic cell. The value of the potential difference of this cell is extremely sensitive to the magnitude of the H_3O^+ concentration and, thereby, to the pH of the solution. In the pH meter-voltmeter, the galvanic cell is connected internally to a voltmeter equipped with a scale that registers pH directly after the instrument has been standardized using solutions of known pH. The instructor will provide you with detailed directions for operating the pH meter used in your laboratory.

Buffer Solutions

Often it is necessary to run a chemical reaction in solution at nearly constant pH value. This can be accomplished by allowing the reaction to occur in the presence of a pH stabilizer called a buffer. A buffer solution has a characteristic pH that changes only slightly when the solution is diluted or when small amounts of strong acid or strong base are introduced. A buffer generally consists of a mixture of an acid-base conjugate pair in some appropriate ratio. Human blood, which must be maintained at about pH 7.4, is buffered by a variety of such conjugate pairs, including (1) dissolved CO_2 and HCO_3^-, (2) $H_2PO_4^-$ and HPO_4^{2-}, and (3) proteins and their conjugate acids.

The pH of a particular buffer solution prepared by mixing an acid-base conjugate pair depends on the value of the acid dissociation constant, K_a (or base dissociation constant, K_b), and on the relative amounts of each component present at equilibrium. For example, an aqueous solution of acetic acid may be mixed with aqueous sodium acetate to form a buffer. The sodium acetate acts as the primary source of acetate ion, the conjugate base of acetic acid. The number of acetate ions formed from the dissociation of the weak acid is usually negligible compared to the number introduced from the strong electrolyte sodium acetate. An expression for the pH characteristic of such a mixture can be derived from the equilibrium expression for K_a of acetic acid. For the acetic acid-acetate system, the equations for the established equilibria are

$$HC_2H_3O_2 + H_2O \rightleftarrows H_3O^+ + C_2H_3O_2^- \tag{17-5}$$

$$K_a = \frac{[H_3O^+][C_2H_3O_2^-]}{[HC_2H_3O_2]} \tag{17-6}$$

Solving eqn. 17-6 for $[H_3O^+]$ and taking the logarithm of each term yields

$$\log[H_3O^+] = \log K_a + \log \frac{[HC_2H_3O_2]}{[C_2H_3O_2^-]} \tag{17-7}$$

The following equivalent expressions for the pH are derived from eqn. 17-7 and eqn. 17-1 and from the definition $pK_a = -\log K_a$.

$$pH \cong pK_a - \log \frac{[HC_2H_3O_2]}{[C_2H_3O_2^-]} = pK_a + \log \frac{[C_2H_3O_2^-]}{[HC_2H_3O_2]} \tag{17-8}$$

Eqn. 17-8 can be used to calculate the pH of an acetic acid-acetate buffer solution when pK_a and the equilibrium concentrations of acetate ion and undissociated acetic acid are known. When these concentrations are equal, pH = pK_a (since log ($[C_2H_3O_2^-]/[HC_2H_3O_2]$) = log 1 = 0. If pK_a is less than 7, the buffer is acidic; if pK_a is greater than 7, the buffer is basic. For acetic acid, $K_a = 1.76 \times 10^{-5}$ at 25°C and $pK_a = 4.75$. Therefore, equimolar concentrations of acetic acid and sodium acetate yield an acidic, buffer of pH 4.75. When these concentrations are not the same, the logarithmic term in eqn. 17-8 will be unequal to 0 and the pH of the buffer will be different from 4.75.

The mechanism of buffer action can be understood by applying Le Chatelier's principle, which states that if stress is imposed on a system at equilibrium, the position of the equilibrium will shift to relieve the stress. For example, if a small amount of strong acid like HCl is added to an acetic acid-acetate buffer, there is an initial surge in the H_3O^+ concentration. The system, however, reacts to reduce this stress by shifting the equilibrium expressed in eqn. 17-5 to the left. During this process most of the added H_3O^+ is consumed by reacting with acetate ions present in the buffer to form undissociated acetic acid. As a result, the original pH of the buffer undergoes a relatively small change. If a small amount of strong base is added, most of the OH^- ions are neutralized when some of the undissociated acetic acid is converted into acetate ion as the equilibrium shifts to the right. Again the pH of the buffer solution remains almost unchanged.

Two major requirements for effective buffer action are: (1) the original concentrations of both acid and conjugate base must be large enough to compensate for the stress to be applied and (2) the logarithm of the ratio of the equilibrium concentrations of acid and conjugate base should be small in magnitude compared to the pK_a. In general, the prepared acid to conjugate base ratio should range between 10^{-1} and 10^1. A given buffer, therefore, has a practical range of about 2 pH units centered at the pK_a value.

An acidic buffer at any desired pH can be prepared by mixing a computed ratio of acid and conjugate base solutions, selected so that K_a of the acid is close to the desired pH. For example, suppose a buffer solution with a pH of 4.92 is to be prepared. Since this pH value is close to the pK_a value for acetic acid, 4.75, the acetic acid-acetate buffer may be chosen. The concentration ratio can be calculated by rearranging eqn. 17-8.

$$\log \frac{[C_2H_3O_2^-]}{[HC_2H_3O_2]} \cong pH - pK_a = 4.92 - 4.75 = 0.17 \tag{17-9}$$

The required $[C_2H_3O_2^-]/[HC_2H_3O_2]$ ratio = antilog 0.17 = 1.48. Therefore, the concentration of acetate ion must be 1.48 times the concentration of acetic acid to produce a buffer of pH 4.92. In making this buffer, the equilibrium concentrations in eqn. 17-9 may be approximated by the respective *original* concentrations of acetic acid and sodium acetate solutions corrected for dilution when the solutions are mixed. (Since acetate is a common ion in eqn. 17-5, the dissociation of acetic acid is repressed in accordance with Le Chatelier's principle.)

Next, consider the following specific example. Prepare 100 mL of a buffer of pH 4.92 from 0.1 M solutions of acetic acid and sodium acetate. How many milliliters of each solution are required? From the definition of molarity, it can be shown that the concentration ratio must be identical to the mole ratio of the buffer components. The *mole ratio* will in turn be equal to the volume ratio.

$$\frac{[C_2H_3O_2^-]}{[HC_2H_3O_2]} \cong \frac{\text{volume of } 0.1\ M\ NaC_2H_3O_2}{\text{volume of } 0.1\ M\ HC_2H_3O_2} \tag{17-10}$$

Since the total volume of buffer solution needed is 100.0 mL, these relations apply:

$$\frac{\text{milliliters of } 0.1\ M\ NaC_2H_3O_2}{\text{milliliters of } 0.1\ M\ HC_2H_3O_2} \cong 1.48 \tag{17-11}$$

$$\text{milliliters of } 0.1\ M\ NaC_2H_3O_2 + \text{milliliters of } 0.1\ M\ HC_2H_3O_2 = 100.0\ \text{mL} \tag{17-12}$$

Let x = milliliters of 0.1 M $H_2C_2H_sO_2$. Substitution of eqn. 17-11 into eqn. 17-12 gives the simple algebraic equation

$$1.48x + x = 100.0 \tag{17-13}$$

Solving eqn. 17-13 for x, we find that the volumes of acetic acid and sodium acetate solutions that must be mixed to prepare 100.0 mL of a buffer with pH 4.92 are:

the volume of $0.1 M$ $HC_2H_3O_2$ = 40.3 mL
the volume of $0.1 M$ $NaC_2H_3O_2$ = 59.7 mL

A buffer solution can also be prepared by mixing a solution of weak base with a solution of the salt (conjugate acid) of that weak base. For example, solutions of ammonia and ammonium chloride may be mixed to form a basic buffer. The pOH of the buffer depends on the value of the base dissociation constant, K_b, and on the ratio of the equilibrium concentration of the two components. At 25° C, the value of K_b for NH = 1.77×10^{-5}. For basic buffers it is often convenient to derive and use an expression analogous to eqn. 17-9 but containing pOH and K_b instead of pH and K_a. For the ammonia-ammonium chloride system, the following relations apply.

$$NH_3 + H_2O \rightleftarrows OH^- + NH_4^+$$

$$K_b = \frac{[NH_4^+][OH^-]}{[NH_3]}$$

$$\log [OH^-] = \log K_b + \log [NH_3]/[NH_4^+]$$

$$\log [NH_4^+]/[NH_3] = pOH - pK_b = pOH - 4.75 \qquad (17\text{-}14)$$

Buffers made from this conjugate pair will be most effective at a pOH approximately equal to 4.75, corresponding to a pH of 9.25.

For example, in the preparation of a buffer of pH 8.60 from ammonia and ammonium chloride, eqn. 17-14 can be used to find the required $[NH_4^+]/[NH_3]$ ratio. Substituting the value pOH = 5.40 into eqn. 17-14 yields

$$\log ([NH_4^+]/[NH_3]) = 5.40 - 4.75 = 0.65$$

$$[NH_4^+]/[NH_3] = 4.47$$

The required volumes of solutions can then be computed as in the example of the acetate-acetic acid buffer.

The ammonia-ammonium chloride system might equally well be considered an acid-conjugate base mixture, where NH_4^+ ion is the acid, and NH_3 the conjugate base. To calculate the $[NH_4^+]/[NH_3]$ ratio, eqn.s similar to 17-8 and 17-9 could be used, where pK_a is the negative logarithm of K_a. Thus, since the pK_a value is greater than 7, a basic solution results from an equimolar mixture of ammonia and ammonium chloride.

PROCEDURE
Wear your safety goggles throughout the experiment.

Indicator Method
You will be assigned one solution from each column of the following list of 0.1 M solutions.

sodium bisulfite	sodium bicarbonate	ammonia
phosphoric acid	boric acid	sodium sulfite
ammonium chloride	zinc chloride	sodium acetate
sodium dihydrogen phosphate	sodium monohydrogen phosphate	sodium carbonate

Measure the pH of each of the three solutions assigned to you. Place about 1 mL of the solution to be tested in a clean 12x75-mm test tube. Wet a strip of Universal indicator paper with a few drops of the solution. (This paper has been impregnated with several indicator dyes.) Determine the approximate pH range by comparing the color you obtain with the colors shown on the side of the paper container. Then, to find a more accurate pH value, select a specific sulfonphthalein indicator that undergoes a color change in that range. Add 1 drop of indicator solution to the test solution and mix well with a clean stirring rod. Compare the resulting color with those shown for that indicator on a color chart. The best way to view the solution is by tilting the test tube at about a 45° angle from the vertical over a white background and looking down through the liquid column (Fig. 17-1). The solution color should correspond to a color in the center of the indicator range. If your solution matches the standard color at either extreme of the indicator range, you should repeat the test with a different indicator, one whose range overlaps the original one.

For each assigned solution record the measured pH value. Write the balanced net ionic equation for a reaction between the assigned reagent and water that can be considered mainly responsible for the observed acidity or basicity.

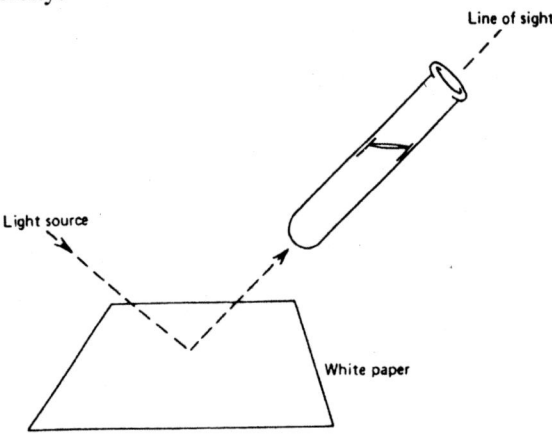

Fig. 17-1. Observing a solution mixed with indicator.

Preparation of a Buffer

Your instructor will assign to you a pH value. Select an appropriate pair of acid-base conjugates from those available and calculate the volumes of 0.1 M solutions of these required for preparation of 100.0 mL of buffer solution at the assigned pH value. Prepare the solution carefully, using a clean 100-mL graduated cylinder. After thorough mixing with a clean glass rod, measure the pH of the solution with a pH meter. (*Consult your instructor before using this delicate instrument.* General information about a pH meter-voltmeter is given in Appendix D.) Record the measured value on the report sheet.

Test for the Effectiveness of the Buffer Solution

Divide the buffer solution prepared above into two 50 mL portions. Add 10 drops of 0.1 M HCl to one portion and 10 drops of 0.1 M NaOH to the other. Stir each of the solutions with a clean glass rod. Then measure the pH value of each with the pH meter and record the values. For comparison, measure and record the pH of two 50 mL portions of distilled water before and after the addition of 10 drops of 0.1 M HCl to one and 10 drops of 0.1 M NaOH to the other.

Determination of K_a or K_b of a Weak Electrolyte from a pH Measurement

Obtain several milliliters of a solution of an unknown acid, HA, or an unknown base, B. Record the prepared molar concentration of the unknown (this information will be provided by your instructor). Measure the pH of the solution with the pH meter or by the indicator method. Calculate the K_a or K_b

depending on whether your unknown is acidic or basic. Use eqn. 17-1 or 17-2 and the appropriate set of generalized equations that follows. (Recall that the concentrations of *all* species in the expressions for K_a or K_b are *equilibrium* concentrations.)

$$HA + H_2O \rightleftarrows H_3O^+ + A^- \qquad\qquad B + H_2O \rightleftarrows BH^+ + OH^-$$

$$K_a = \frac{[H_3O^+][A^-]}{[HA]} \qquad\qquad K_b = \frac{[BH^+][OH^-]}{[B]}$$

FURTHER READING
1. https://fscimage.fishersci.com/msds/95544.htm (0.1 N HCl, last accessed July, 2019)
2. http://www.labchem.com/tools/msds/msds/LC24270.pdf (0.1 N NaOH, last accessed July, 2019)
3. https://www.spectrumchemical.com/MSDS/LC11080.pdf (0.1 N NH$_3$, last accessed July, 2019)
4. http://www.labchem.com/tools/msds/msds/LC10972.pdf (0.1 N NH$_4$Cl, last accessed July, 2019)
5. http://www.labchem.com/tools/msds/msds/LC10380.pdf (0.1 N HC$_2$H$_3$O$_2$, last accessed July, 2019)
6. https://fscimage.fishersci.com/msds/20860.htm (NaC$_2$H$_3$O$_2$, last accessed July, 2019)
7. http://www.labchem.com/tools/msds/msds/LC11715.pdf (H$_3$BO$_3$, last accessed July, 2019)
8. http://www.labchem.com/tools/msds/msds/LC18640.pdf (H$_3$PO$_4$, last accessed July, 2019)
9. http://www.labchem.com/tools/msds/msds/LC24775.pdf NaH$_2$PO$_4$, last accessed July, 2019)
10. http://www.labchem.com/tools/msds/msds/LC24774.pdf (Na$_2$HPO$_4$, last accessed July, 2019)
11. https://fscimage.fishersci.com/msds/21080.htm (Na$_2$CO$_3$, last accessed July, 2019)
12. https://fscimage.fishersci.com/msds/20970.htm (NaHCO$_3$, last accessed July, 2019)
13. https://fscimage.fishersci.com/msds/88494.htm (Na$_2$SO$_3$, last accessed July, 2019)
14. https://www.spectrumchemical.com/MSDS/S3700.pdf (NaHSO$_3$, last accessed July, 2019)
15. https://fscimage.fishersci.com/msds/25350.htm (ZnCl$_2$, last accessed July, 2019)
16. Stock, J. T. "The pH Indicator" *The Science Teacher* **29** (1962) 28
17. Strong, C. L. *Sci. Amer.* **219** (1968) 232
18. Zuehlke, R. W. *J. Chem. Educ.* **39** (1962) 354

COMMENTS

Name _____ Lab Section _____ Date _____   193

Prelaboratory Assignment: Experiment 17
pH Measurement and Buffer Solutions

Where appropriate, answers should be given to the correct number of significant digits.

1. Define a buffer solution. How does it differ from a solution of a strong base?

2. What is the definition of K_w, K_a, and K_b? In aqueous solution, what is the relationship between these constants? What two species are involved in the calculation of K_a and K_b [acid and conjugate base]?

3. Show your calculations to receive full credit. What is the pH of the following aqueous solutions?
   a. 3M HCl

   b. 10M HCl

   c. 10^{-3}M HCl

   d. 2×10^{-8} M HCl

4. Describe at least two different methods for determining a solution pH.

5. Based on the Henderson-Hasslebalch equation, will pH = pK_a be equivalent if the acid and conjugate base concentration are not equivalent?

6. Under what conditions does pOH = pK_b?

Experiment 17 pH Measurement and Buffer Solutions

Name _____ Lab Section _____ Date _____ 195

REPORT ON EXPERIMENT 17
pH Measurement and Buffer Solutions
Where appropriate, answers should be given to the correct number of significant digits.

DATA AND RESULTS

Indicator Method

Name of Solution	pH	Net Ionic Equation
1. _____	_____	_____
2. _____	_____	_____
3. _____	_____	_____

Preparation of a Buffer

    assigned pH   _____

    reagents used   _____

    mL of conjugate acid used   _____

    mL of conjugate base used   _____

    measured pH   _____

Test for the Effectiveness of the Buffer Solution

Substance	Acid or Base Added	Measured pH
50 mL of buffer	10 drops of 0.1 M HCl	_____
50 mL of buffer	10 drops of 0.1 M NaOH	_____
water	none	_____
50 mL of water	10 drops of 0.1 M HCl	_____
50 mL of water	10 drops of 0.1 M NaOH	_____

Experiment 17 pH Measurement and Buffer Solutions

Determination of K_a or K_b of a Weak Electrolyte from a pH Measurement

method used to measure pH _____

unknown number _____

molarity of unknown _____

pH _____

$[H_3O^+]$ _____

molarity of unknown at equilibrium _____

K_a (if acidic) _____

pOH _____

$[OH^-]$ _____

K_b (if basic) _____

Show the details of all calculations; use extra sheets if necessary.

QUESTIONS

(Submit your answers on a separate sheet as necessary.)

1.
   a. What was the change in pH (ΔpH) of your buffer upon adding 10 drops of HCl? ΔpH = pH(final) – pH(initial).

   b. What was the ΔpH for pure water upon adding 10 drops of HCl?

   c. Explain why one of these changes is of a magnitude larger than/the same as the other.

2. Suppose in the experiment instead of adding 10 drops of HCl you added 50 mL of 0.1 M HCl to your buffer. Would you expect your buffer to remain effective? Why or why not?

Experiment 17 pH Measurement and Buffer Solutions

3. Suppose that after you prepared your 50 mL of buffer, your lab partner accidentally added 10 mL of pure water to it. How would that affect its pH? Explain.

4. When determining the Ka (or Kb) of a substance in the last part of this experiment, suppose that your lab partner accidentally added 10mL of pure water to it just before the pH measurement. How would that affect its pH? Explain.

5. Under what circumstances is the pH of a solution negative? Give one example of such a solution.

COMMENTS

Experiment 18
Natural Acid-Base Indicators and pH

OBJECTIVE

To investigate the chemical and physical properties of natural indicators derived from plant sources. To determine the pH of several solutions.

EQUIPMENT

Safety goggles, gloves, 24-well micro plate, 12 × 75-mm test tubes, 25 × 200 mm test tubes, Genesys 20 or similar spectrophotometer, cuvettes, Denver Ultra-basic UB-5 or similar pH meter, pH paper, 50-mL Erlenmeyer flask, (*optional*) Whatman No. 4 filter paper. OPTIONAL EQUIPMENT: Absorbance Microplate Reader, 24-well glass or acrylic plate, centrifuge, ≤ 2mm mesh sieve, 4-dram amber vial.

REAGENTS

200 mL vinegar, 200 mL 0.1M HCl, 200 mL 0.1M NaOH, 200 mL 0.1M NH$_4$OH, 200 mL 0.1M NaCl, 200 mL deionized water, 200 mL 0.1M KHP, and 200 mL 0.1 M KOH, red sorghum natural indicator, phenolphthalein, methyl orange and/or methyl red. **NOTE**: The natural indicator will be extracted by the lab instructor.

SAFETY AND DISPOSAL

Refer to the MSDS information available online when working with solutions of HCl,[1] and NaOH,[2] NH$_4$OH,[3] NaCl,[4] KHP,[5] KOH,[6] Ethanol (absolute)[7], methyl red[8] phenolphthalein[9] and methyl orange.[10]

Disposal for these compounds and their derived salts should be in accordance with local, state and federal regulations. All indicators have both organic and inorganic elements (Cl$^-$, F$^-$) in them after reaction with salts, acids and bases. Dispose of these wastes in a container labeled for halogenated compounds. Disposal for unreacted acids should be into a labeled laboratory waste jar for acids. Disposal for unreacted bases should be into a labeled laboratory waste jar for bases.

INTRODUCTION

Indicators are dyes used to assist with accurately determining the pH of any solution. Many synthetic indicators and dyes are in commercial use. However, some of these dyes and indicators are expensive and have been classified as a detriment to the ecosystem, especially the food chain. In this experiment, we will use what we classify as a *natural indicator*. A natural indicator is a compound found in nature that contains pigments that can be used to detect the equivalence point of an acid base titration. These natural indicators can come from various sources such as leaves, tree bark or fruits and vegetables. These sources contain a group of organic compounds known as *flavonoids* (*anthocyanin* or *anthocyanindins*). Some examples of natural indicator sources are cabbage leaves, *sorghum bicolor* red leaves, Hibiscus plants, and berries.

Extraction of these indicators using solvents such as ethanol or water makes this a green or environmentally friendly experiment. Some natural indicators have the ability to cause a color change even when the pH shift from acid to base or vice versa is minimal. These indicators will be used along with commercially available synthetic indicators to determine the pH of several different solutions.

pH is measured by taking the negative of the logarithm of the concentration of hydrogen ions in a particular solution and is expressed in Eqn. 18-1:

$$pH = -\log [H^+]. \tag{18-1}$$

Acid-base indicators are usually weak Brønsted acids or bases. If the indicator is a Brønsted acid (an H^+ donor), it will produce a conjugate base. If the indicator is a Brønsted base (an H^+ acceptor), it will produce a conjugate acid. Many times, we study acids and bases using a process called titration which we covered in Experiment 8. When the acid-base reaction reaches the equivalence point, this means that the moles of the acid and base are equal. This is not the endpoint however. To determine the endpoint, we need to use an indicator. The endpoint in an acid base reaction is the point in which the indicator causes the solution to change colors. The behavior of the indicator is due to the color difference between the acid and its conjugate base. The equilibrium between the acidic form (H-Indicator) and the conjugate base form (Indicator⁻) in aqueous solution might be expressed as shown in Fig. 18-1:

$$\text{H-Indicator} + H_2O \rightleftharpoons H_3O^+ + \text{Indicator}^-$$
Brønsted Acid (Yellow)   Conjugate Base (Pink)

Fig. 18-1. Acidic indicator in aqueous solution

Based on LeChatelier's principle, one can predict whether any given indicator will be present in its acidic or basic form. At lower pH ranges (*i.e.*, more H^+ present), LeChatelier's principle dictates that the reaction should move in the reverse direction to maintain the equilibrium thereby making a pink solution turn yellow. Based on Fig. 18-1, if the indicator is acidic and the solution is acidic, there should be an even more intense yellow color because we have increased the H+ concentration.

Adding base to the solution containing the acid and the acidic indicator will increase the concentration of hydroxide ions (OH⁻) and concomitantly decrease the hydrogen ion (H^+) concentration. This will cause a color shift from the yellow (acidic pH region) to pink (basic pH region). The specific color change described above is observed with *Sorghum bicolor* indicator. Sorghum bicolor, extracted from red sorghum leaves, is acidic and causes a yellow color in the presence of an acid. Other natural indicators display different colors depending on their nature (acidic or basic) and the nature of the solution to which they are added (acidic or basic).

There are other indicators that show a progression of color change based on pH. If an indicator is *multi-protic* (which means that it can be deprotonated more than once), it will display multiple colors as the pH shifts from acidic to neutral. When these indicators are used in a titration and the acid base reaction reaches the endpoint, the color remains constant. It is therefore important to note that the specific pH ranges where color changes occur depend greatly on the equilibrium between the conjugate acid and base forms of the indicator.

Chemistry of Natural indicators

The color changes discussed above are attributed to the presence of different functional groups on the indicator. In the case of *Apigenin or apigenidinin*, that can be obtained from strawberries, blueberries or sorghum seeds, hydroxyl groups that are present are able to donate protons (Fig. 18-2).[11]

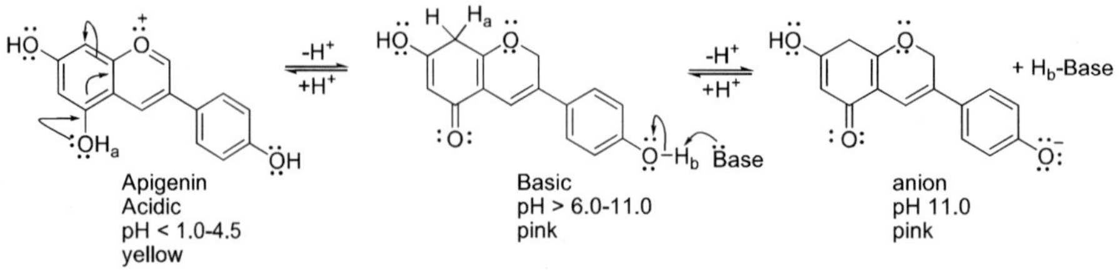

Apigenin
Acidic
pH < 1.0-4.5
yellow

Basic
pH > 6.0-11.0
pink

anion
pH 11.0
pink

Fig. 18-2. Mechanism for apigenin protonation/deprotonation

PROCEDURE

Part A. Measuring changes in absorbance and pH based on the addition of indicator
a. Obtain a 50-mL Erlenmeyer flask.
b. From your instructor obtain 20 mL of plant extract that has anthocyanin/anthocyanidin containing pigments (natural indicator).
c. Record the color of the pigment in Table 1 of your data sheet.
d. Add 2 mL of the extract to a cuvette and determine λ_{max} by measuring the absorbance of the solution using a spectrophotometer.
e. Measure the pH of the natural indicator extracted using pH paper or a pH meter and record your findings in Table 1 of the data sheet. Was the indicator that you used acidic or basic? How could you tell?

Part B. Qualitative/Quantitative testing of indicators
a. Obtain 14 test tubes (12 × 75 mm) from your instructor.
b. To 7 test tubes, add 2 mL each of 0.1 M NaOH, 0.1 M HCl, 0.1M KHP, 0.1M NH$_4$OH, 0.1M NaCl, vinegar, and distilled water
c. Measure the pH of the solution in each test tube.
d. Add two to three drops of your natural indicator extract into each test tube and shake carefully to mix.
e. Let the test tubes stand for two minutes. Record any observations in your lab sheet.
f. Repeat steps 1 and 4 using two of the following: phenolphthalein, methyl orange or methyl red as your indicator. Record your observations in your data sheet.

Part C. Microtitration of acids and bases
a. Select 6 test tubes and label three as "natural indicator" and the remaining three as phenolphthalein.
b. Add 2.0 mL of 0.1M HCl acid into each of the test tubes.
c. Measure the pH of the solution in each test tube.
d. Add 2-3 drops of the selected natural indicator (see Part A) to the first three tubes labeled "natural indicator".
e. Measure the pH of the three test tubes labeled "natural indicator".
f. Add 2-3 drops of phenolphthalein indicator to the three test tubes labeled phenolphthalein.
g. Measure the pH of the three test tubes labeled phenolphthalein.
h. To the first test tube labeled natural indicator, add 1 mL increments of 0.1M NaOH until a color change is observed. Record the amount of base added. Repeat for 0.1 M NH$_4$OH and 0.1 M KOH in test tubes 2 and 3 labeled, "natural indicator."

Optional Extraction Procedure for apigenin indicator solution[13]
Weigh approximately 1.00 g of a powdered sample leaves (≤ 2 mm mesh) into a Pyrex culture test tube (25 × 200 mm) and add 25.0 mL of ethanol (99.9%). The mixture should be centrifuged for 5 minutes at ambient temperature (25°C) and then filtered using Whatman No. 4 filter paper into a new culture test tube, capped and ready for use on the same day. **Caution:** The powder should be sieved (≤ 2mm mesh) into an amber bottle and stored away from direct sunlight to prevent photolysis and decomposition prior to preparation of the indicator.

Optional Procedure for Determining Absorbance with a Plate Reader
a. Obtain a 24-well plate from your instructor.
b. Follow the template below and add 1 mL KHP to each well in row 1, add 1 mL of 0.1M HCl to each well in row 2, add 1 mL of 0.1M NH$_4$OH to each well in row 3, add 1 mL of 0.1M NaCl to each well in row 4, add 1 mL NaOH to each well in row 5 and add 1 mL vinegar to each well in row 6.

c. Choose two natural indicators.
d. Add two drops of your first natural indicator extract across lane 1.
e. Add two drops of your second natural indicator extract across lane 2.
f. Add two drops of phenolphthalein to each well across lane 3.
g. Add two drops of methyl red to each well across lane 4.
h. Allow the solutions to mix for five minutes. Record any observations in your lab sheet.
i. **[OPTIONAL EXERCISE]** Using a plate reader, record the absorbance for each well. Are the results different than the results that you observed using your spectrophotometer?

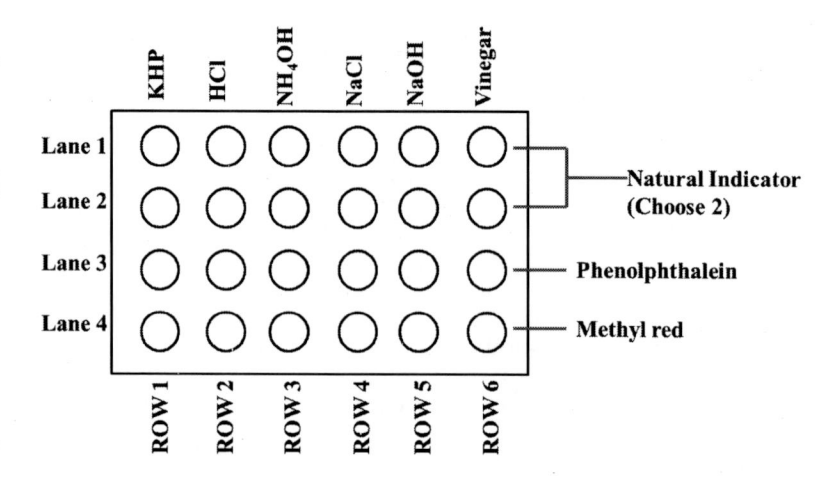

FURTHER READING
1. http://www.labchem.com/tools/msds/msds/LC15320.pdf (HCl, last accessed July, 2019)
2. http://www.labchem.com/tools/msds/msds/LC24350.pdf (NaOH, last accessed July, 2019)
3. https://fscimage.fishersci.com/msds/00211.htm (NH$_4$OH, last accessed July, 2019)
4. https://fscimage.fishersci.com/msds/21105.htm (NaCl, last accessed July, 2019)
5. https://fscimage.fishersci.com/msds/19425.htm (KHP, last accessed July, 2019)
6. http://www.labchem.com/tools/msds/msds/LC19190.pdf (KOH, last accessed July, 2019)
7. https://www.fishersci.com/shop/msdsproxy?storeId=10652&productName=BP28184 (200 proof alcohol, last accessed November, 2013)
8. https://fscimage.fishersci.com/msds/45435.htm (Methyl red indicator)
9. https://fscimage.fishersci.com/msds/96382.htm (Phenolphthalein indicator)
10. https://fscimage.fishersci.com/msds/60355.htm (Methyl orange indicator)
11. Mendham, J.; Denney, R.C.; Barnes, J.D.; Thomas, M.J.K. *Textbook of Quantitative Chemical Analysis,* 6th ed.; Pearson Education Limited: U.K., **2000**
12. Abugri, D. A.; Pritchett, [G]; Apea, O.B.; Russell, A.E.; Akudago, J.A.; Abugri, J.B.; Tay-Agbozo, S.S. *"A Novel Cheaper Natural Indicator: Proposed Mechanisms and Indicator-effectiveness Properties of Selected Solvent Extracts"*, private communication
13. Abugri, D. A.; Apea, OB.; and Pritchett G. "Investigation of a Simple and Cheap Source of a Natural Indicator for Acid-Base Titration: Effects of System Conditions on Natural Indicators" *Green and Sustainable Chemistry*, (2012), **2**, 117-122.

Name _____ Lab Section _____ Date _____

Prelaboratory Assignment: Experiment 18
Natural Acid-Base Indicators and pH

1. State three applications of natural dyes?

2. What are the three (3) forms of natural dyes?

3. State three (3) ways to separate natural dyes?

Experiment 18 Natural Acid-Base Indicators

4. Explain the terms green chemistry and sustainability from a natural dye standpoint?

5. What will be the charge of a natural indicator (*Sorghum bicolor*) that is in a basic medium?

Name _____ Lab Section _____ Date _____

REPORT ON EXPERIMENT 18
Natural Acid-Base Indicators

DATA SHEET

Part A. Measuring changes in absorbance and pH based on the addition of indicator

Indicator Name	Color	pH	Absorbance	Nature of Indicator (acidic or basic)

Part B. Qualitative/Quantitative testing of indicators

Indicator Name	Solution	Color change indicator added	pH of solution with indicator	pH of solution w/o indicator
	0.1 M HCl			
	0.1 M NaCl			
	0.1 M NaOH			
	Vinegar			
	Distilled water			
	0.1 M NH$_4$OH			
	0.1 M KHP			
	0.1 M HCl			
	0.1 M NaCl			
	0.1 M NaOH			
	Vinegar			
	Distilled water			
	0.1 M NH$_4$OH			
	0.1 M KHP			

Experiment 18 Natural Acid-Base Indicators

Part C. Microtitration of HCl using different bases

Indicator	Solution	Solution pH w/o indicator	Solution pH w/ indicator	Color w/base added	Amount of base added for color change
Natural	0.1 M HCl/ 0.1 M NaOH				
Natural	0.1 M HCl/ 0.1 M NH$_4$OH				
Natural	0.1 M HCl/ 0.1 M KOH				
Phenolphthalein	0.1 M HCl/ 0.1 M NaOH				
Phenolphthalein	0.1 M HCl/ 0.1 M NH$_4$OH				
Phenolphthalein	0.1 M HCl/ 0.1 M KOH				

QUESTIONS *(Submit your answers on a separate sheet as necessary.)*

1. State at least three (3) principles of green chemistry and sustainability that can be explained using the natural indicator you studied in this laboratory?

2. If the natural indicator was added to an acid, what color change will you observe? Explain.

3. If the natural indicator was added to urine what color change do you expect to occur? Explain.

4. If HCl reacted with KOH in a simple titration, using the Sorghum indicator, do you expect the mole ratio to remain in 1:1? Explain.

5. Both phenolphthalein and *Sorghum bicolor* indicator signal an end point of an acid-base reaction with a pink color. If you were to add these indicators independently to a hemoglobin solution. What will the color be for each indicator in the solution of hemoglobin?

Experiment 19
Solubility Product

OBJECTIVE

To determine the solubility of calcium sulfate in water by using ion exchange techniques; to obtain an approximation to the solubility product of calcium sulfate.

EQUIPMENT

30- to 50-cm glass column (approximately 1-cm i.d.) with one end drawn out and joined to a 4- to 5-cm glass nozzle using 4 to 5 cm of rubber tubing with a Hoffman screw clamp (a 25- or 50-mL buret may be used as a glass column), 7.5-cm short stem funnel, glass wool, 30 to 50-cm glass rod (5- to 7-mm o.d.), 250-mL Erlenmeyer flask, 100-mL volumetric flask, 250-mL beaker, 100-mL graduated cylinder, 10-mL transfer pipet, buret stand, clamp and fastener, pH meter, electrode assembly, Universal pH paper, filter paper.

REAGENTS

2 g calcium sulfate, 10 g Dowex-50(x 8) 50- to 100-mesh resin, standard pH 4 buffer solution, 25 mL 4M HCl.

SAFETY AND DISPOSAL

Refer to the MSDS information available online when working with $CaSO_4$,[1] HCl[2]. Disposal for these compounds and their derived salts should be in accordance with local, state and federal regulations. Disposal for $CaSO_4$ should be into a labeled laboratory waste container for inorganic chemicals. Disposal for unreacted acids should be into a labeled laboratory waste jar for acids.

INTRODUCTION

In this experiment an ion exchange resin will be used to determine the amount Ca^{2+} in solution when $CaSO_4$ is dissolved in water to form an aqueous solution. This information, along with the amount of "free" or undissociated $CaSO_4$, will be used to calculate the solubility product for $CaSO_4$. Ion exchange resins are small beaded solid polymeric materials containing sites of either positive or negative charge to which are attached small counter ions of opposite charge. When an aqueous solution of an electrolyte is brought into contact with an ion exchange resin, exchange between the counter ions and the electrolyte takes place. For example, a resin, RH, which has sites R^- and counter ions H^+ immersed in a sodium chloride solution might have the following equilibrium occur:

$$RH \text{ (s)} + Na^+ \rightleftarrows H^+ + RNa$$

The resin has exchanged H^+ ions for Na^+ ions from the solution. Hence, the resin is a cation exchanger. Similarly, resins containing positively charged sites and small soluble anions are called anion exchangers. Since a balance of charge must be maintained in the solution and in the resin, equivalent quantities of charge must be maintained. When Ca^{2+} is put in contact with RH, the following reaction occurs:

$$2RH \text{ (s)} + Ca^{2+} \rightleftarrows R_2Ca \text{ (s)} + 2H^+$$

A solution of any salt can be *quantitatively* converted to an acidic solution by having it react with a cation exchange resin (or to a basic solution by using an anion exchanger). To achieve efficient exchange, the salt solution is passed through a column of ion exchange resin. As the solution moves down the column, it continuously contacts layer of the resin that have not yet reacted. If the column is sufficiently long, the effluent (the solution leaving the column) does not contain any of the original salt cations. The solubility product describes a relationship between the activities of ions in equilibrium with the solid compound. In dilute solution, concentrations of ions provide an approximate measure of their activities. Therefore, the solubility product also describes an approximate relationship between concentrations of ions in a saturated solution of a compound. Consider a solid AB that yields ions A^+ and B^- in solution. When such a solution is in equilibrium with solid AB, the product of the molar concentrations of ions A^+ and B^- is constant at a given temperature and gives a reasonable estimate of the solubility product constant:

$$K_{sp} = [A^+][B^-]. \tag{19-1}$$

The square brackets denote molar concentrations. For a strong electrolyte, A_xB_y the solubility product takes the form:

$$K_{sp} = [A^{y+}]^x[B^{x-}]^y. \tag{19-2}$$

It is calculated by raising the molar concentrations of the product ions to the power of the number of these ions produced in solution by the dissociation of a molecule of A_xB_y. For example, one Ba^{2+} ion and two F^- ions are formed when a molecule of BaF_2 dissolves. The solubility product is given by

$$K_{sp} = [Ba^{2+}][F^-]^2. \tag{19-3}$$

In a pure solution of an ionizable compound, the concentrations of the cation and the anion are related. In the example of BaF_2, if the concentration of Ba^{2+} ion in a saturated solution of the salt were determined to be 7.52×10^{-3} M, then the F^- ion concentration must be $2 \times 7.52 \times 10^{-3}$ M. The solubility product would equal:

$$K_{sp} = (7.52 \times 10^{-3})(2 \times 7.52 \times 10^{-3})^2 = 1.70 \times 10^{-6}.$$

In saturated solutions of slightly soluble salts, the component ions are present in such small concentrations that their determination requires sensitive methods of analysis. Moreover, the relationship between solubility and solubility product can be complicated by ion pair formation. Ion pairs are formed when small, highly charged cations combine with anions to yield aqueous solvated complexes. For example, in aqueous solutions of $CaSO_4$ the following reaction occurs:

$$Ca^{2+} + SO_4^{2-} \rightleftarrows CaSO_4\,(aq). \tag{19-4}$$

Ion pair formation is particularly significant when the magnitude of the product of the cation charge and the anion charge is greater than 2. Thus, for solid $CaSO_4$, an equilibrium is established between ions and ion pairs such that:

$$CaSO_4(s) \rightleftarrows Ca^{2+} + SO_4^{2-} \rightleftarrows CaSO_4\,(aq). \tag{19-5}$$

In a saturated solution of $CaSO_4$ at 25°C, approximately 60% of the ions present are in the form of ion pairs. Hence, the solubility product obtained by direct calculation of the solubility of $CaSO_4$ is much higher than the actual product of the concentrations of the Ca^{2+} and SO_4^{2-} ions. The extent of ion pair formation can be calculated from the thermodynamic dissociation constant of $CaSO_4\,(aq)$, which is given by

$$\frac{[Ca^{2+}][SO_4^{2-}]}{[CaSO_4(aq)]} = 5.2 \times 10^{-3}. \tag{19-6}$$

If the solubility of the CaSO$_4$ is known, then a good estimate of the concentrations CaSO$_4$ (aq), Ca^{2+}, and SO$_4^{2-}$ can be obtained. Consider the following example.

y = solubility of CaSO$_4$ (s) in moles per liter
c = concentration of Ca^{2+} ions

In the reaction
$$CaSO_4\,(aq) \rightleftarrows Ca^{2+} + SO_4^{2-}$$

for every Ca^{2+} ion formed, one SO$_4^{2-}$ ion is formed. Therefore,

c = concentration of SO$_4^{2-}$ = concentration of Ca^{2+} ions
$y - c$ = concentration of CaSO$_4$ (aq)

Substitution into eqn. 19-6 yields
$$\frac{c \times c}{y - c} = 5.2 \times 10^{-3}$$

$$c^2 = 5.2 \times 10^{-3}(y-c)$$

Solving for c using the quadratic formula gives

$$c = \frac{-5.2 \times 10^{-3} \pm (2.7 \times 10^{-5} + 2.08 \times 10^{-2}y)^{0.5}}{2}$$

Thus, a good estimate of the concentration of Ca^{2+} and SO$_4^{2-}$ can be made by measuring y, the solubility of CaSO$_4$. A reasonable approximation (Note 19-1) of the solubility product can be obtained using the value of c and the relationship

$$K_{sp} = c^2 \qquad (19\text{-}7)$$

In this experiment, the solubility of calcium sulfate is evaluated by determining the Ca^{2+} ion concentration in a saturated solution of the salt. The analytical technique used is ion exchange, which is simple and inexpensive. An aliquot of a saturated solution of calcium sulfate is passed through an acidified cation exchange column. The Ca^{2+} ions in solution are taken up (or sorbed) on the solid ion exchange resin in the column, and H$^+$ ions are displaced from the exchanger into the solution according to the reaction:

$$2HR + Ca^{2+} \rightleftarrows CaR_2 + 2H^+.$$

Ion pair formation within the aliquot is destroyed as Ca^{2+} ions are exchanged for H$^+$ ions. Two moles of H$^+$ ions will be exchanged for every mole of Ca^{2+} ions present in the solution coming out of the column (the effluent). A pH meter can then measure the concentration of acid in the effluent. The solubility of CaSO$_4$, (the y value), and thereby a good estimate of the solubility product, (the value of c^2) can then be calculated.

PROCEDURE

Wear your safety goggles throughout the experiment. Review the material on gravity filtration, the transfer of liquid samples, and the use of the transfer pipet in the Laboratory Equipment and Techniques section before attempting this procedure.

Using a digital balance, measure 2g of calcium sulfate into a 250-mL beaker. Add about 100 mL of deionized or distilled water. While stirring, heat the mixture to 80°C and then allow it to cool to room temperature.

Preparation of an Ion Exchange Column

If possible, prepare your ion exchange column the period before this experiment is done. Otherwise, assemble an ion exchange column while the $CaSO_4$ solution is cooling. Use either a 25- or 50-mL buret, or a 30- to 50-cm glass column (1-cm inner diameter) with one end drawn out, and attach it to a glass nozzle (4- to 5-mm outer diameter) by means of a 4- to 5-cm piece of rubber tubing to construct the apparatus shown in Fig. 19-1. The rate of flow of solution through a buret can be controlled by its stopcock, while the rate of flow through the 30- to 50-cm glass column can be controlled by Hoffman screw clamp. Fill the buret or glass column with distilled water. No air bubbles should remain in the nozzle. Make a small ball of glass wool of roughly the same diameter as the glass column, wet it with water, and insert it in the bottom of the column by means of a long glass rod with a flattened end. Place a clean beaker under the column to catch any effluent.

On a digital balance weigh about 10 g of Dowex-50 (x8) cation exchange resin in a 150-mL beaker. Make a slurry by adding 30 mL of distilled water, and pour it through a funnel into the column. Keep several centimeters of water always above the level of the exchange resin (Note 19-2). To make more room in the column, open the Hoffman screw clamp and let some of the water run out. After all the resin has been introduced into the column (use more water if necessary), lower the water level to about 1 cm above the resin bed.

Regeneration of the Column

If the resin being used here has not been previously prepared or is not in the acid form, or if it has been used before this experiment, it must be brought into the acid form. This is done by passing through the column 25 mL of 4 M HCl solution at a flow rate of approximately 1 mL per minute. Next, wash the column by passing water through it at as fast a flow rate as possible, until Universal pH paper indicates it to be of the same pH as distilled water. Discard all the effluent collected.

Loading and Washing the Column

When the saturated $CaSO_4$ solution has cooled to room temperature, filter it by gravity filtration into a clean 250-mL Erlenmeyer flask to obtain a clear solution. Transfer 5 to 10 mL of the saturated solution into the filter. Guide the solution cleanly from the beaker into the funnel with a glass rod. Use the first few milliliters of the filtrate to rinse the Erlenmeyer flask. Remove the funnel temporarily, swirl the solution about the inner walls of the flask, and discard the rinse solution. Continue the filtration until you have a clear, saturated solution of calcium sulfate. Record the temperature of this solution on the report sheet.

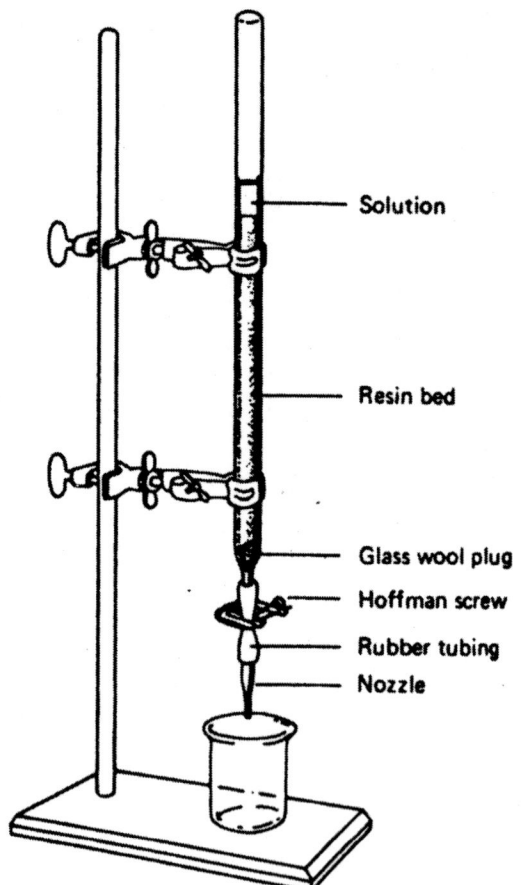

Fig. 19-1 Ion Exchange Column

Bring the level of water in the column to a few milliliters above the level of the resin bed. Pipet precisely 10.00 mL of the filtered solution into the ion exchange column. Place a clean 100-mL volumetric flask under the column. Adjust the flow rate on the column to about 0.5 mL per minute or 8 to 10 drops per minute. Pipet an additional 20 mL of the saturated solution (in two precise 10-mL portions)

Experiment 19 Solubility Product

into the column whenever there is enough room. Stop the column when the level of the solution inside it has fallen to only 4 mm or so above the resin bed. Introduce about 10 mL of distilled water to wash the column.

Collect this washing in the same volumetric flask at the same flow rate. Again, stop the column when the level of the solution is barely above the resin bed. Now introduce 20 mL of distilled water into the column and let it flow into the same volumetric flask but at a faster rate.

Test a drop of the effluent with Universal pH paper toward the end of this washing step. If the pH indicated is not significantly different from that of distilled water, stop the washing and dilute the effluent in the volumetric flask with distilled water to the graduation mark. If the effluent is still acidic, continue the washing until it is not acidic or until the 100-mL volumetric flask has been filled to the graduation mark (Note 19-3).

Standardize a pH meter, using standard buffer solution as described by your instructor. If you are using a pH meter-voltmeter, see Appendix D for general information about a pH meter. Determine the pH of the diluted effluent solution by transferring 50 mL of it to a clean, pre-rinsed 150-mL beaker. Introduce the electrode assembly into this solution and read the pH. Noting that the H_3O^+ ion concentration is twice that of Ca^{2+} and that 30.0 mL of saturated solution was put through the column, calculate the solubility of $CaSO_4$. Then calculate the solubility product of $CaSO_4$, using eqn. 19-7.

NOTES
19-1   The error introduced by using concentrations instead of activities to calculate K_{sp} is described in detail in the article by Meites *et al.* See *Further Reading*.
19-2   If the water level falls below the top of the resin bed, air enters the column and channels are created which allow the liquid phase to bypass some of the resin; therefore, complete ion exchange does not occur.
19-3   In the latter case, only a negligible amount of the exchange acid is left behind in the column.

FURTHER READING
1. https://esciencelabs.com/sites/default/files/msds_files/Calcium%20Sulfate.pdf ($CaSO_4$, last accessed July, 2019)
2. http://www.labchem.com/tools/msds/msds/LC15300.pdf (HCl, last accessed July, 2019)
3. Koubek, D. "The Solubility of $CaSO_4$: An Ion Exchange Complexometric Titration Experiment for the Freshman Laboratory." *J. Chem. Educ.* **53** (1976) 254
4. Meites, L.; Pode, J.S.F.; and Thomas, H.C. "Are Solubilities and Solubility Products Related?" *J. Chem. Educ.* **43** (1966) 667

COMMENTS

Prelaboratory Assignment:
Experiment 19 Solubility Product
Where appropriate, answers should be given to the correct number of significant digits.

1. Formation of ion pairs in solution (raises / lowers) the molar solubility of the solid relative to what is predicted by K_{sp}. Explain your choice below.

2. Write the chemical equation for the K_{sp} of $CaSO_4$.

3. Write the ion pair dissociation reaction for $CaSO_4$.

4. What types of compounds have a tendency to form ion pairs in solution?

Experiment 19 Solubility Product

5. In this experiment, why is it necessary to heat a solution of calcium sulfate also containing solid calcium sulfate?

6. Why is it necessary to keep at least 4 mm of water above the resin column at all times?

Experiment 19 Solubility Product

Name _____ Lab Section _____ Date _____ 215

REPORT ON EXPERIMENT 19
Solubility Product
Where appropriate, answers should be given to the correct number of significant digits.

DATA AND RESULTS

    temperature   _____

    volume of saturated solution passing through column   _____

    volume of effluent after dilution   _____

    pH of effluent   _____

    molarity of H⁺ ion in effluent   _____

    solubility of $CaSO_4$ in saturated solution   _____

    solubility product, K_{sp} _____

Show the details of all calculations; use extra sheets as necessary.

QUESTIONS
(Submit your answers on a separate sheet as necessary.)

1. Calculate the molarity of a saturated solution of CaF_2 at 18°C (K_{sp} = 3.40 × 10⁻¹¹). Assume that only Ca^{2+} and F^- are present in solution.

2. A 50.0-mL sample of a saturated CaF_2 solution at 18° C is passed through an ion exchange column in OH^- form. The effluent is diluted to 100 mL with distilled water. Calculate the molarity of $Ca(OH)_2$ in this solution.

Experiment 19 Solubility Product

3. Calculate the number of moles of OH⁻ ions per liter of the solution in question 2 (first and second dissociation constants are 3.74×10^{-3} and 4.00×10^{-2}, respectively, for $Ca(OH)_2$).

4. Calculate the pH of the effluent solution in question 2.

Experiment 19 Solubility Product

Experiment 20
Standard State Determinations for Urea Solvation

OBJECTIVE
To determine ΔH° and K for the dissolution of urea in water and thereby calculate standard state values ΔG° and ΔS°.

EQUIPMENT
Thermometer (alcohol or digital, 0°C – 100°C), stirring rod, two 250-mL (8 oz.) Styrofoam cups, one-hole no. 2-rubber stopper, ring stand, Buret clamp, 10-mL graduated cylinder, 25-mL beaker, electronic balance.

REAGENTS
Urea, water.

SAFETY AND DISPOSAL
Refer to the MSDS information available online when working with urea.[2] Urea is an irritant. Wear gloves when handling it.

INTRODUCTION[1]
Urea, shown in Fig. 20-1, is an organic compound formed by the body after metabolizing protein. The average person excretes about 30 grams of urea daily, mostly through urine, with a small amount as perspiration. A good approximation to the enthalpy of the solvation for urea in water will be determined through calorimetry.

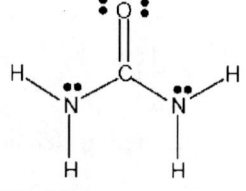

Fig. 20-1 Urea

Calorimetry
When a chemical is introduced into the calorimeter that is initially at room temperature and a process takes place, both the calorimeter and the chemical will change temperature. A property of the system (the calorimeter and sample) called the *change in enthalpy* is defined as the heat absorbed or emitted in a reaction taking place at constant pressure with no other work than pressure-volume work occurring. The change in enthalpy is often referred to as the change in "heat content" for a constant temperature process. The change in enthalpy of the sample must be the negative of the change in enthalpy for the calorimeter. However, the calorimeter parts (the calorimeter vessel, thermometer, stirrer and other materials (besides water) that may be present) absorb some of the heat. To correct for heat lost to the calorimeter parts, a correction factor is used. This correction factor is called the *calorimeter heat capacity*, C_{pc}. (See note 20-1).

Consider an example. A 5.0 g sulfuric acid sample at 23°C is added to 40.0 mL of water in a calorimeter at 23.0°C. The final temperature of the mixture is 45.0°C. Assume the density of liquid water to be 1.00 g mL^{-1} at all temperatures. The calorimeter heat capacity is 18.8 J deg^{-1}. Determine the enthalpy of solvation for sulfuric acid.

$$\text{change in enthalpy of solution} + \text{change in enthalpy of calorimeter} = 0$$

	change in enthalpy	+	change in enthalpy	+	change in enthalpy		
	of solution		of calorimeter water		of calorimeter system	=	0

$$\Delta H + m\, c_{p1}\, \Delta T + C_{pc}\, \Delta T = 0 \quad \quad (20\text{-}1)$$

where m = mass of water
c_{p1} = specific heat of water in J g^{-1}deg^{-1}
C_{pc} = calorimeter heat capacity
ΔT = temperature change in °C, always expressed as $T_{final} - T_{initial}$.

Substitution of the example values in equation 20-1 yields:

$$\Delta H_{solvation} + (40.0\text{g})(4.184\text{ J g}^{-1}\text{deg}^{-1})(45.0-23.0)\text{deg} + (18.8\text{ J deg}^{-1})(45.0-23.0)\text{deg} = 0$$

$$\Delta H_{solvation} = -(3.68\text{ kJ} + 414\text{ J}) = -4.09\text{ kJ}$$

We can use this type of calculation to determine $\Delta H_{solvation}$ for urea. Next, we will make an approximation that the observed $\Delta H_{solvation} = \Delta H°$.

The equilibrium constant K will be determined by measuring the water solubility of urea. The equilibrium expression is given by:

$$\text{urea }(s) + \text{water} \rightleftarrows \text{urea }(aq)$$

Therefore,

$$K = [\text{urea}] \quad \quad (20\text{-}2)$$

Thus, the equilibrium constant $K = [\text{urea}]$. With the enthalpy change and the equilibrium constant determined experimentally, the free energy change and enthalpy change for the process are readily calculated:

$$\Delta G° = -RT \ln K \quad \quad (20\text{-}3)$$

To determine $\Delta S°$, the standard state entropy, we can then use the relationship that

$$\Delta G° = \Delta H° + T\Delta S° \quad \quad (20\text{-}4)$$

and your observed values for $\Delta G°$, $\Delta H°$, and of course T.

PROCEDURE:
Wear your safety goggles throughout the experiment.

Part 1. Determination of ΔH (ΔH°)
Create a calorimeter by nesting two Styrofoam cups. Add 50.0 mL of water to the calorimeter. Use a Buret clamp, a one-hole rubber stopper, and a ring stand, to mount a thermometer so that its bulb is well into the water in the calorimeter. Record the initial temperature of the water. Accurately weigh and record the weight of 2-3 g of urea. *Rapidly* add the urea to the water with *good mixing*. Make sure that all of the urea dissolves relatively quickly in order to get the best temperature measurement. Read and record the final temperature of the solution *immediately* after the urea has completely dissolved. Repeat this procedure two more times for a total of three trials.

For each trial, calculate and record the heat of dissolution, as shown in the sample calculation above. Since the concentration of urea is low, you can assume the heat capacity of the solution is essentially that of pure water. Use 18.8 J deg^{-1} for the calorimeter heat capacity. Use the average value of the three trials for your final value of $\Delta H°$. Alternatively, you can make the temperature measurements with a digital thermometer. Your instructor will provide instructions for the use of a digital thermometer.

Part 2. Determination of K
Weigh and record the weight of 3-5 g of urea and place it in a 25-mL beaker. Slowly add 2-3 mL of water with good mixing. Then proceed to add water drop-wise after that, stirring sufficiently between each addition, until the urea has just completely dissolved. After the urea has just dissolved, record the temperature of the solution. Then remove the thermometer, pour the solution into a 10-mL graduated cylinder and record the final volume of the solution. Repeat this procedure two more times for a total of three trials. For each trial, record the volume and calculate the value of the equilibrium constant.

After you have calculated ΔH and K, and assume that $\Delta H = \Delta H°$, you can calculate $\Delta G°$ and $\Delta S°$ for the process. Use Eqns. 20-2, 20-3, and 20-4 in your calculations.

NOTES
20-1. There is a potential source of error in determining the maximum temperature reached by the calorimeter. As the calorimeter temperature rises after the introduction of the urea sample, the calorimeter is losing heat to its surroundings. Thus, the maximum observed temperature change is somewhat less that the actual temperature change. When highly precise measurements are required, such losses are corrected for by reading the temperature of the calorimeter at regular time intervals before, during and after the addition of urea. A plot of temperature against time is made, and the rate of heat loss is determined from the graph by extrapolation to zero time. However, it is not necessary to correct for such heat losses in this experiment. With the simple apparatus used, the error from this source is less than the error in measuring the temperature change.

FURTHER READING
1. Liberko, C. A.; Terry, S. *J. Chem. Educ.* 2001, *78*, 1087
2. https://fscimage.fishersci.com/msds/24680.htm (urea, last accessed July, 2019)
3. Armstrong, G. T.; *J. Chem. Educ.* **41** (1964) 297
4. Wilhoit, R. C.; *J. Chem. Educ.* **44** (1967) A571
5. Zaslow, B. J.; *J. Chem. Educ.* **37** (1960) 578

COMMENTS

Name _____ Lab Section _____ Date _____

Prelaboratory Assignment: Experiment 20
Standard State Determinations for Urea Solvation

Where appropriate, answers should be given to the correct number of significant digits.

1. You add 2.50 g of urea to 6.00 mL of water. The urea dissolves.
   a. Is the reaction urea(s) → urea(*aq*) spontaneous? Justify your choice.

   b. If the reaction is spontaneous, which of the following must be true?
      i. $\Delta G° > 0$ $\Delta H° > 0$ $\Delta S° > 0$
      ii. $\Delta G° < 0$ $\Delta H° < 0$ $\Delta S° < 0$

   c. Justify your answer to question 1b.

   d. What experimental evidence would you obtain if a reaction were endothermic?

   e. If the reaction were endothermic, which of the following must be true?
      i. $\Delta G° > 0$
      ii. $\Delta H° > 0$
      iii. $\Delta S° > 0$
      iv. $\Delta G° < 0$
      v. $\Delta H° < 0$
      vi. $\Delta S° < 0$

   f. Justify your answer to question 1e.

2. Using the internet or your textbook, determine the relationship between *K* and the concentration of a saturated urea solution.

3. Do you expect that dissolving a crystalline solid in water to form a solution will result in higher or lower entropy? Explain your hypothesis.

Experiment 20 Standard State Determination for Urea Solvation

4. Given that 2.50 g of urea dissolve in 6.00 mL of water with $\Delta T = 4$ °C, how mathematically will you obtain the values for:

   a. K for the solvation of urea in water

   b. $\Delta H°$

   c. $\Delta G°$

   d. $\Delta S°$

Name _____ Lab Section _____ Date _____ 223

REPORT ON EXPERIMENT 20
Measuring the Entropy Change of Urea Solvation
Where appropriate, answers should be given to the correct number of significant digits.

Part 1. Determination of ΔH (ΔH°)	Trial 1	Trial 2	Trial 3
Weight of Urea (grams)	_____	_____	_____
Initial Temperature (°C)	_____	_____	_____
Final Temperature (°C)	_____	_____	_____
Volume of water (mL)	_____	_____	_____
Weight of solution (grams)	_____	_____	_____
ΔH° (in kJ/mol)	_____	_____	_____
Average ΔH° (in kJ/mol)		_____	

Part 2. Determination of K			
Weight of Urea (in grams)	_____	_____	_____
Final Temperature (in °C)	_____	_____	_____
Final Volume (in mL)	_____	_____	_____
Concentration of Urea (mol/L)	_____	_____	_____
K	_____	_____	_____
Average K		_____	
ΔG° (in kJ/mol)	_____	_____	_____
Average ΔG° (in kJ/mol)		_____	
ΔS° (in kJ/mol-K)	_____	_____	_____
Average ΔS° (in kJ/mol-K)		_____	

Experiment 20 Standard State Determination for Urea Solvation

QUESTIONS

1. Based upon your results, what is the saturation concentration of urea? Explain.

2. From your experimental results, is dissolving urea spontaneous? Explain.

3. Is dissolution of urea exothermic or endothermic? Explain.

4. Do your results for $\Delta S°$ support your hypothesis about the change in entropy for dissolving a crystalline substance from the prelab questions? Explain.

5. Since $\Delta G° = \Delta H° - T\Delta S°$, reactions are often analyzed as to whether their behavior is dominated by enthalpy (the first term on the right) or by entropy (the second term). Which of these terms dominates the urea dissolution process?

Experiment 20 Standard State Determination for Urea Solvation

Experiment 21
Galvanic Cells

OBJECTIVE

To construct a half-cell and measure its reduction potential; to determine, using chemical analysis, the direction of current through an external circuit; and to observe the effect of concentration on the cell potential.

EQUIPMENT

150-mm drying tube; no. 0 one-hole rubber stopper; 5-mm o.d. glass tubing; voltmeter or pH meter-potentiometer; glass microscope slides; rubber bands; cotton balls; ring stand and extension buret clamps; zinc, copper, aluminum, graphite, and lead electrodes; 0.10 g no. 20 mesh or small strips of electrode materials; 250-mL beaker; medicine dropper; steel wool or sandpaper; no. 22 or similar diameter copper wire; permeable plastic membrane or Saran wrap; 18 X 150-mm test tube.

REAGENTS

0.10 g no. 20 mesh lead shot; 20-mL viscous solution of agar saturated with $NaNO_3$; 30-mL of 1 M $Pb(NO_3)_2$; 25mL each per constructed cell of the following solutions: 1 M $ZnSO_4$, 1 M $CuSO_4$, 2 M $FeSO_4$, 1 M KI that is 0.08 M in I_2, 1 M $Al(NO_3)_3$.

SAFETY AND DISPOSAL

Refer to the MSDS information available online when working with lead shot,[1] $NaNO_3$,[2] $Pb(NO_3)_2$,[3] $ZnSO_4$,[4] $CuSO_4$,[5] $FeSO_4$,[6] KI[7] that is 0.08 M in I_2,[8] $Al(NO_3)_3$.[9] Disposal for these compounds and their derived salts should be in accordance with local, state and federal regulations. Disposal for these materials should be into a labeled laboratory waste container for inorganic chemicals.

INTRODUCTION

Galvanic cells are devices in which chemical energy is converted into electrical energy by an oxidation-reduction (redox) reaction. Such a reaction occurs when electrons provided by one reagent, R (the reducing agent), are accepted by another reagent, Ox (the oxidizing agent). The process may be considered as two half reactions, each called a redox couple.

$$\text{oxidation half reaction:} \quad R \rightarrow R^{n+} + ne^- \quad \quad (21\text{-}1)$$

$$\text{reduction half reaction:} \quad Ox + me^- \rightarrow Ox^{m-} \quad \quad (21\text{-}2)$$

$$\text{overall reaction:} \quad mR + nOx \rightarrow mR^{n+} + nOx^{m-} \quad \quad (21\text{-}3)$$

Various galvanic cells can be constructed by choosing different combinations of R and Ox. The observed electrical potential for the overall reaction may be regarded as the sum of contributions from each half reaction.

When the respective reagents are separated so that they are in electrical contact but do not react directly with each other, electrons may be transferred between them through an external conductor. The flow of electrons through the conductor constitutes an electric current and is evidence of a difference in electrical potential between the two redox couples. Each redox couple, along with the electrode that connects it to the external circuit, is called a *half-cell*. To construct a galvanic cell, contact is made between the two half-cells by a salt bridge or by a porous partition. The salt bridge or the porous partition each permit the passage of ions but inhibit direct chemical reaction.

The electrode where oxidation takes place is called the *anode*. Ions that migrate toward the *anode* are called *anions*. The electrode at which reduction takes place is called the *cathode*. Ions that migrate toward the *cathode* are called *cations*.

Line Notation
A convenient system for schematically representing electrochemical cells has been developed. The rules are:
1. The anode is written on the left, the cathode on the right.
2. Vertical bars represent phase boundaries (e.g., the boundaries between solid electrode and solution).
3. Double vertical bars represent a salt bridge.
4. Broken vertical lines represent a liquid junction (*e.g.*, a porous partition).
5. The solutes in each solution are listed in any convenient order and their concentrations are given in parentheses after the formulas.

For example, a Daniell cell (a special type of galvanic cell) may consist of zinc and copper half-cells connected by a salt bridge. Each half-cell consists of a metal electrode partially immersed in a 1.00 M aqueous solution of its metal ion. The electrode reactions are:

anode: $Zn \rightarrow Zn^{2+} + 2e^-$

cathode: $Cu^{2+} + 2e^- \rightarrow Cu$

overall: $Cu^{2+} + Zn \rightarrow Zn^{2+} + Cu$

Using line notation, the cell may be represented by
$$Zn \mid Zn^{2+} (1.00\ M) \parallel Cu^{2+} (1.00\ M) \mid Cu$$

If a porous partition is used instead of the salt bridge, the line notation is written
$$Zn \mid Zn^{2+} (1.00\ M) : Cu^{2+} (1.00\ M) \mid Cu$$

Standard Electrode Potential
The electrical potential produced in a spontaneous redox reaction and the direction of the current depend on the relative reducing or oxidizing strength of the chemical substances, on their concentrations, the temperature and on the solution medium. A given redox couple will have a specific potential to donate or accept electrons. However, it is possible to measure only the difference in potential between the two half-cells. Hence, a numerical scale of half-cell potentials is obtained by arbitrarily defining as zero potential the potential of a half-cell that consists of hydrogen gas in contact with a solution of hydrogen ions and in which all reagents are in their standard states at unit activity.

Experimentally, this standard state may be approximated by bubbling hydrogen gas at 1.00 atm pressure over a platinum electrode in contact with a 1.00 molal (1.00 m) solution of hydrogen ions furnished from an *n-n* electrolyte (note 21-1). An *n-n* electrolyte is one that furnishes n cations and n anions in solution. For example, HCl and NaCl are 1-1 electrolytes; they furnish one cation for each anion in solution. $CuCl_2$ is an example of a 1-2 electrolyte because one cation is furnished for every two anions.

Since the standard electrode potential, ε^o, of the hydrogen half-cell is defined as zero, the potential of any other half-cell is merely the observed potential difference when it is part of a cell which also includes a standard hydrogen half-cell. The standard electrode potential is obtained when the reagents are in their standard states at unit activity. For solutions of *n-n* electrolytes, the standard state is represented by a hypothetical 1.00 m ideal solution that is often approximated by a 1.00 m real solution.

The sign and magnitude of a standard reduction electrode potential, ε^O, may be determined in the following manner. A galvanic cell is constructed consisting of a standard hydrogen half-cell and the standard half-cell of unknown potential. The potential difference of the complete cell is then measured. If the hydrogen half-cell is acting as the anode (an electron transmitter), the other half-cell is functioning as the cathode (an electron receiver). The standard reduction potential of the latter is then *positive* and equal in magnitude to the observed potential difference.

If, on the other hand, the hydrogen half-cell is acting as the cathode in the galvanic cell, the other half-cell functions as the anode with a positive standard oxidation potential equal to the observed potential difference. The standard *reduction* potential (which is always opposite in sign to the standard oxidation potential) is equal in this case to the observed cell potential difference with a *negative* sign. You will find a list of standard reduction potentials in Appendix B, Table 8. Those standard half-cells with negative ε^O values will lose electrons to hydrogen half-cells under standard conditions. Those with positive ε^O values will receive electrons from the hydrogen half-cells.

It is generally inconvenient to use standard hydrogen half-cells in the determination of electrode potentials. Instead, any half-cell with an accurately known standard potential may be employed as a reference electrode in a galvanic cell. If the reference electrode functions as the *anode* in the cell, the observed cell potential difference, $\Delta\varepsilon^O$, is given by:

$$\Delta\varepsilon^O = \varepsilon^O - \varepsilon^O_{ref} \tag{21-4}$$

where ε^O and ε^O_{ref} are the standard reduction potentials of the unknown and reference electrodes, respectively.

When the reference electrode functions as the *cathode*:

$$\Delta\varepsilon^O = \varepsilon^O_{ref} - \varepsilon^O \tag{21-5}$$

Thus, ε^O can be determined by measuring $\Delta\varepsilon^O$, establishing the direction of electron flow, and using eqn. 21-1 or 21-2. For example, in a Daniell cell, the observed potential difference between the standard zinc and copper electrodes is $\Delta\varepsilon^O = 1.10$ V (volts). The Cu I Cu^{2+} couple, which acts as the cathode here, is known to have a standard reduction potential of $\varepsilon^O = +0.34$ V. Using eqn. 21-2 we find that the Zn I Zn^{2+} couple has a standard reduction potential of $\varepsilon^O = -0.76$ V.

In this experiment, you will construct a half-cell from materials designated by your instructor. The electrolyte used will consist of 1 M aqueous solution (instead of the 1 m solutions used in defining half-cell potentials). Also, the electrolyte may not be of the *n-n* type (for example, $Al(NO_3)_3$ solution is not *n-n*). For the purposes of this experiment, however, the results of such approximations are sufficiently precise to use in calculations and determinations. The half-cell potential can be determined by connecting your half-cell to a reference Pb I Pb^{2+} half-cell and measuring the cell potential. The reference half-cell has a standard reduction potential of $\varepsilon^O = -0.13$ V. If the direction of the current is such that the Pb IPb^{2+} couple is the anode, the ε^O for your half-cell may be calculated from eqn. 21-1. If the Pb I Pb^{2+} functions as the cathode, eqn. 21-2 may be used.

A galvanic cell will then be constructed by combining two of the given half-cells (yours and one constructed by another student). The direction of current through an external circuit will be determined by chemical methods of analysis. This is done by using the galvanic cell as a source of current to electrolyze an aqueous solution.

Depending upon the choice of electrode, the electrolysis reactions are:

cathode: $2H_2O + 2e^- \rightarrow H_2(g) + 2OH^-$
anode: $Cu \rightarrow Cu^{2+} + 2e^-$
or
$2H_2O \rightarrow O_2(g) + 4H^+ + 4e^-$

There is an increase in the OH^- concentration at the electrode at which electrons enter the aqueous solution (leave the galvanic cell). Thus, by using an indicator that detects OH^-, you can determine the direction of the current through an external circuit. The cathode in the electrolytic cell receives electrons from the anode in the galvanic cell.

Experiment 21 Galvanic Cells

Effect of Concentration

The electrochemical potential, E, developed in a galvanic cell is also a function of the temperature and activities of the reacting species. If a galvanic cell undergoes the following redox reaction

$$aA + bB \rightarrow cC + dD$$

in which n moles of electrons are transferred between oxidizing and reducing agents, E may be expressed by the Nernst equation, which is approximated by the following form at 298 K and 1.00 atm pressure.

$$\varepsilon = \Delta\varepsilon^O - \frac{0.0592V}{n} \log \frac{[C]^c[D]^d}{[A]^a[B]^b} \tag{21-6}$$

Here, $\Delta\varepsilon^O$ is the difference in standard-state potential of the oxidation and reduction half reactions. Brackets enclosing the symbols of the reagents denote their concentration in moles per liter.

Thus, it is possible to obtain work from an electrochemical cell in which the half-cells contain identical molecular species at different concentrations. For example, consider the cell

$$Cu \mid Cu^{2+} (0.0010 \ M) \parallel Cu^{2+} (0.50 \ M) \mid Cu$$

The cell reaction and potential are determined as follows.

Cell Reaction

anode: $Cu \rightarrow Cu^{2+} (0.0010 \ M) + 2e^-$

cathode: $Cu^{2+} (0.50 \ M) + 2e^- \rightarrow Cu$

overall: $Cu^{2+} (0.50 \ M) \rightarrow Cu^{2+} (0.0010 \ M)$

Cell Potential

$$\varepsilon = 0.0V - \frac{0.0592 \ V}{n} \log \frac{0.0010}{0.50}$$

$$= -0.0296 \ (-2.7) \ V$$

$$\varepsilon = +0.080 \ V$$

The positive value of the cell potential (+ 0.080 V) indicates that the cell proceeds spontaneously as written. Thus, the cell potential developed by a concentration cell may be derived from the Nernst equation to yield the relationship:

$$\varepsilon = -\frac{0.0592 \ V}{n} \log \frac{c_2}{c_1} \tag{21-7}$$

where c_2 is the concentration of the product, c_1 is the concentration of the reactant, and n is the number of moles of electrons transferred as written in the cell reaction. The more concentrated solution spontaneously becomes more dilute, and the more dilute solution becomes more concentrated. When half-cell concentrations are equal, the cell potential is zero. Thus, the potential generated in a concentration cell is expected to decrease as the cell is used and the half-cell solutions become closer in concentration. A standard voltmeter can be used to acquire a qualitative understanding of the concentration cell. Quantitative work requires the use of a potentiometer, standardized solutions, and more exacting experimental conditions.

PROCEDURE
Wear your safety goggles throughout the experiment.

Construction of a Cell
Your instructor will designate the type of half-cell to be constructed. Fit a 150-mm drying tube with a one-hole no. 0 rubber stopper. Fire-polish both ends of a 10-cm piece of 5-mm outer diameter glass tubing. After the glass has cooled, insert it into the rubber stopper.

Preparation of a salt bridge. Pack a small wad of cotton loosely in the narrow tip of the drying tube. Dip the tip into a hot solution of agar-$NaNO_3$ (prepared by your instructor) so that the liquid rises above the cotton wad. This tube serves as the salt bridge shown in Fig.21-1. Remove the tube and set it aside to cool for at least 10 minutes. While waiting for the agar to congeal, perform the experiment on spontaneous cell reactions under *Determination of the Half-Cell Potential*.

Preparation of a porous membrane. The salt bridge can be replaced with a porous membrane. A porous membrane allows ion transport without actually mixing the cell solutions. It can be easily constructed by wrapping the end of the cell in Saran wrap or cellophane. Secure the membrane to the drying tube with rubber bands. Make sure that there are no air bubbles in the drying tube nipple.

Once either the salt bridge or the porous membrane has been constructed, the experimental procedure is the same. Prepare the required electrode according to the following instructions.

Zinc. Clean the surface by dipping the metal briefly into dilute HCl and then rinsing with water.

Copper. Clean with sandpaper.

Aluminum. Just before use, soak in a beaker of 6 M HCl until bubbles of hydrogen appear all over the surface. Do not rinse. Use immediately, before the oxide coating reforms.

Fe^{3+}, Fe^{2+} or I_2, I^-. Clean the surface of a graphite rod with sandpaper and attach a 25-cm piece of copper wire to one end.

After the agar has congealed in the stem of the drying tube, clamp the tube vertically on a ring stand and half-fill it with a 1 M solution which is required to complete the assigned redox couple. (The I_2,I^- solution contains iodine dissolved in a 1 M solution of KI. For the Fe^{3+}, Fe^{2+} couple, use equal volumes of 2 M iron (III) and 2 M iron (II) solutions.) Next, attach a 25-cm piece of copper wire to one end of the electrode and insert the other end of the copper wire through glass tubing in the stopper. Adjust the wire's length so that when it is inserted into the drying tube, the electrode dips into the solution but the copper wire lead does not. Coil the copper wire about the outside of the glass tube so that it does not slide up and down. Complete the half-cell by inserting the stopper into the drying tube.

Determination of the Half-Cell Potential
Spontaneous cell reactions. Measure about 5 mL of 1 M $Pb(NO_3)_2$ solution into an 18x150-mm test tube and add a small strip of material made of the same metal as the assigned electrode. Wait 2 to 3 minutes and note whether a spontaneous deposition of lead occurs. If the assigned electrode is either the I_2,I^- or the Fe^{3+}, Fe^{2+} couple, add 1 to 2 mL of the electrolytic solution to the $Pb(NO_3)_2$ solution. See if a spontaneous deposition of lead occurs.) Dispose of these materials in the appropriate waste jar.

If you observe no reaction, repeat the procedure using about 2 mL of 1 M metal ion (M^{n+}), or electrolytic solution and two or three lead pellets. Notice whether a spontaneous deposition of the metal (M) occurs.

From your experimental observations, deduce which of the following reactions is spontaneous.

$$2M + n\, Pb^{2+} \rightarrow 2M^{n+} + n\, Pb \qquad (21\text{-}8)$$
$$n\, Pb + 2M^{n+} \rightarrow 2M + n\, Pb^{2+} \qquad (21\text{-}9)$$

Determine which electrode serves as the anode and which as the cathode in the galvanic cell. If the agar salt bridge is being used, complete construction of the assigned half-cell and then proceed to the next part of the experiment.

Measurement of the galvanic cell potential.
The half-cell potential will be determined using a voltmeter. (Your instructor will give you detailed instructions for using the voltmeter.) After rinsing the tip of the half-cell with a little 1*M* lead nitrate solution, mount the half-cell in the beaker containing the Pb | Pb^{2+} electrode as shown in Fig. 21-2. The voltmeter will register a positive voltage when its positive terminal is connected to the anode (the electrode where electrons leave the galvanic cell).

Reverse the connections if this is not the case. Make sure all wires make good electrical contact and record the voltage reading. Note which half-cell functions as the anode and find the "standard" potential of your assigned half-cell from eqn. 21-1 or 21-2. Report the electrode potential to your instructor, who will list the results of the entire class. Copy these data and arrange them in an experimental series of reduction potentials.

Optional Calculations. Review the material on uncertainty and error analysis in the Mathematical Treatment of Data section. Then compute:

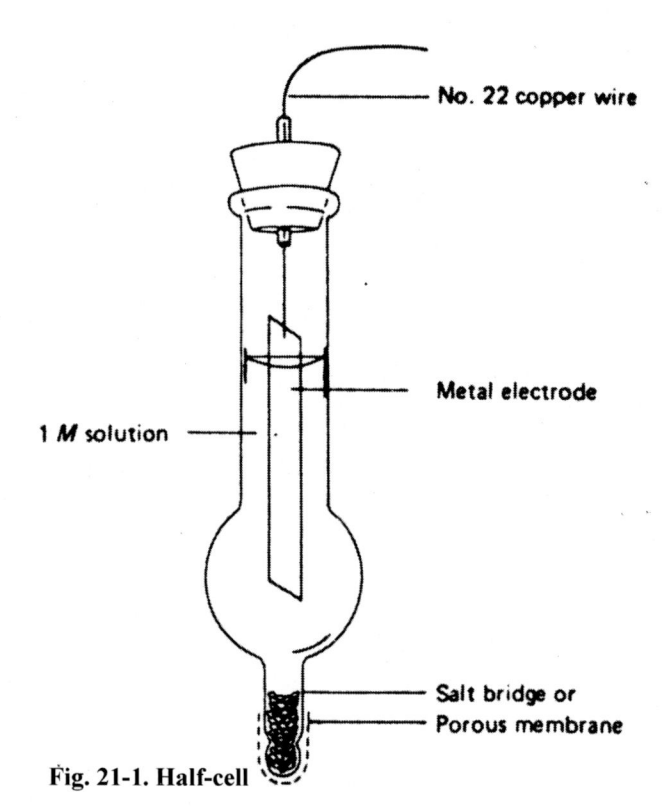

Fig. 21-1. Half-cell

1. the mean value of the "standard" reduction potential for each series of galvanic cells used by the class.
2. the deviation of your value from the mean of the series in which it is included
3. the standard deviation of the mean value of the series in which your cell is included
4. the 95% confidence level of your value for the "standard" reduction potential.

Determination of the Direction of Current by Chemical Analysis

Work with another student who has a chemically different half-cell and construct a galvanic cell according to the following instructions.

Pour 15 mL of saturated $NaNO_3$ solution into a 250-mL beaker. Using a ring stand and a Buret clamp, mount the half-cells in the beaker as shown in Fig. 21-3. Dip a piece of universal indicator paper into the $NaNO_3$ saturated solution and place it on a microscope slide. Connect the two half-cells by setting the wires across the paper about 4 to 5 cm apart. Cover the paper with a second slide and clamp the slides together with a rubber band. Within 5 to 10 minutes, there should be a color change in at least one region where the wire touches the paper (Note 21-2). From the color change, deduce the direction of electron flow in the cell and identify the anode and the cathode. Make several new cells by combining different half-cells and discover the directions of their currents.

Effect of Concentration

Construct two identical half-cells as described earlier. Connect the two half-cells and measure the cell potential. Calculate the expected value of ε for this cell.

Dilute one of the half-cells by a factor of 10. Calculate the cell potential, using eqn. (21-4). Measure and record the new cell potential. Dilute the same cell by another factor of 10 and measure and record the cell potential. Recalculate its potential and compare the computed value to the observed value. Return the electrodes to the stockroom and dispose of all electrolytic solutions used in this experiment in waste jars specially set up for that purpose or as directed by your instructor.

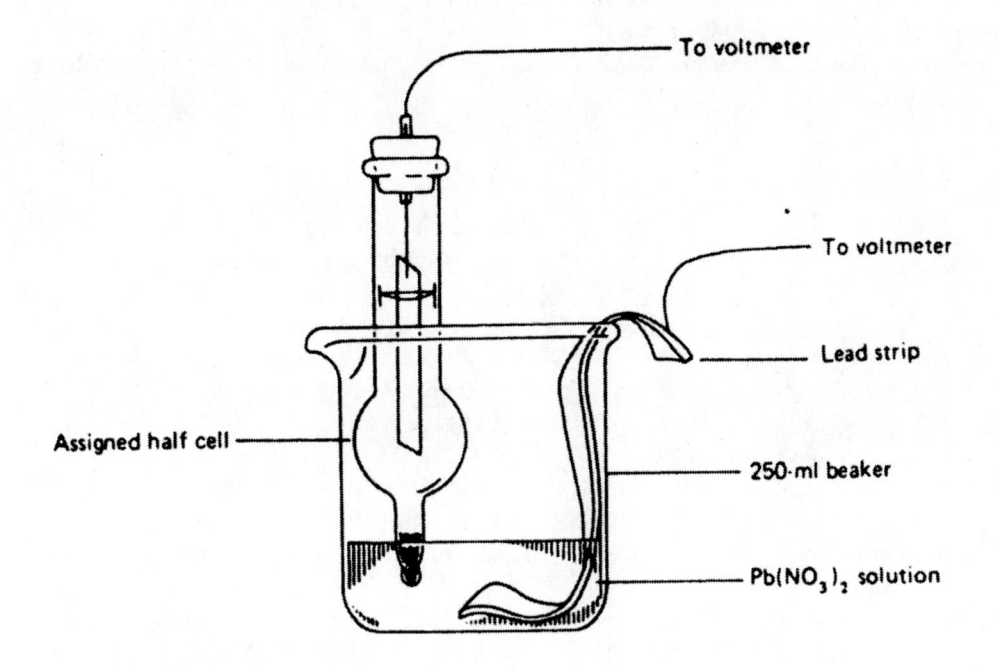

Fig. 21-2. Measurement of galvanic cell potential. Supporting clamps not shown.

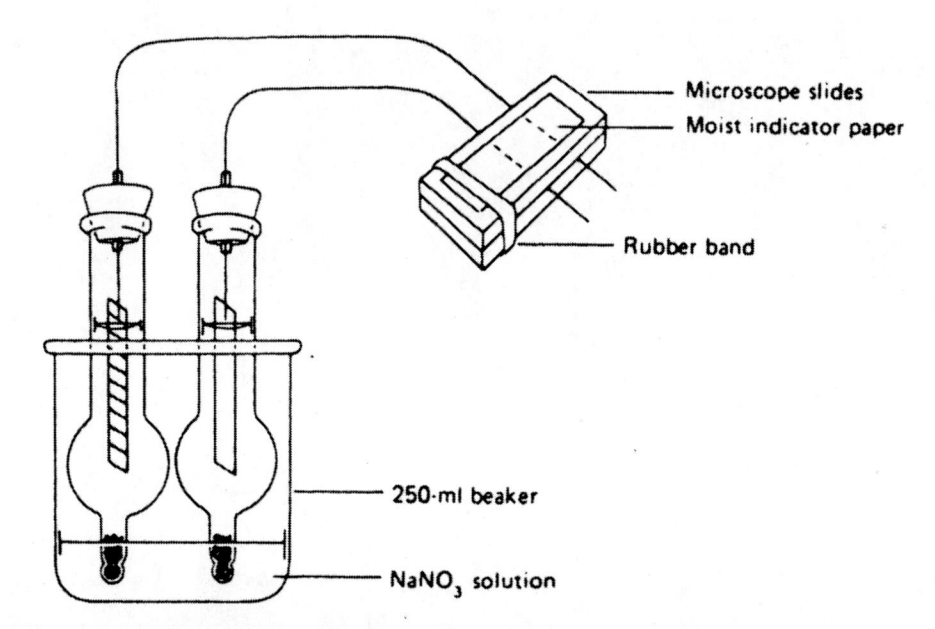

Fig. 21-3. Assembled galvanic cell. Supporting clamps not shown.

Experiment 21 Galvanic Cells

NOTES

21-1. The molality of a solute is the number of moles of solute per kilogram of solvent. It is a temperature-independent unit of concentration.

21-2. When the cell potential difference is small (0.2 V or less), it may take much longer for the color to appear.

FURTHER READING

1. https://fscimage.fishersci.com/msds/12510.htm (Pb, last accessed July, 2019)
2. https://www.fishersci.com/store/msds?partNumber=S343500&productDescription=SOD+NITRATE+ACS+500G&vendorId=VN00033897&countryCode=US&language=en ($NaNO_3$, last accessed July, 2019)
3. https://fscimage.fishersci.com/msds/12660.htm ($Pb(NO_3)_2$, last accessed July, 2019)
4. https://fscimage.fishersci.com/msds/25580.htm ($ZnSO_4$, last accessed July, 2019)
5. https://fscimage.fishersci.com/msds/05690.htm ($CuSO_4$, last accessed July, 2019)
6. https://fscimage.fishersci.com/msds/09870.htm ($FeSO_4$, last accessed July, 2019)
7. http://www.labchem.com/tools/msds/msds/LC19690.pdf (KI, last accessed July, 2019)
8. https://fscimage.fishersci.com/msds/11400.htm (I_2, last accessed November, 2013)
9. https://fscimage.fishersci.com/msds/00942.htm ($Al(NO_3)_3$, last accessed July, 2019)
10. Dillard, C.R., and Kammeyer, P. "An Experiment with Galvanic Cells for the General Laboratory." *J. Chem. Educ.* **40** (1963) 363.
11. Gorman, M. "Some Electrochemical Experiments for Freshman." *J. Chem. Educ.* **34** (1957) 409.
12. Lauren, P.M. "The Dynamics of an Electrolysis." *The Science Teacher* **32** (1965) 60.
13. Robbins, O., Jr. "The Proper Definition of Standard Electromotive Force, Textbook Errors" *J. Chem. Educ.* **48** (1971) 737.
14. Sanderson, R.T. "On the Significance of Electrode Potentials." *J. Chem. Educ.* **43** (1966) 584.
15. Slabaugh, W.H. "Corrosion." *J. Chem. Educ.* **61** (1974) 218.

Name _____ Lab Section _____ Date _____   233

Prelaboratory Assignment: Experiment 21
Galvanic Cells

1. What is the function of a salt bridge? Does its function differ significantly from a porous membrane?

2. Give three examples of *n-n* electrolytes.

3. Where each metal electrode is immersed in a 1 M aqueous solution of its metal ion, write the line notation for a galvanic cell in which the following reaction occurs: $Pb^{2+} + Zn \rightarrow Zn^{2+} + Pb$

Experiment 21 Galvanic Cells

4. Draw a picture of the cell described above in question 3. Clearly indicate the direction of electron flow, the anode, the cathode, and the salt bridge.

5. As found on a materials safety data sheet (MSDS), list the physical and chemical properties of any ONE of the following: lead shot, $NaNO_3$, $Pb(NO_3)_2$, $ZnSO_4$, $CdSO_4$, $CuSO_4$, $FeSO_4$, KI, I_2, $Al(NO_3)_3$. Be sure to give the source of your MSDS information.

Name _____ Lab Section _____ Date _____ 235

REPORT ON EXPERIMENT 21
Galvanic Cells
Where appropriate, answers should be given to the correct number of significant digits.

DATA AND RESULTS

assigned half-cell _____

Determination of the Half-Cell Potential

with respect to the Pb/Pb^{2+} electrode, the
unknown half-cell was (anode, cathode) _____

observed potential _____

"standard" reduction potential _____

class-determined reduction potential series:

Half-Cell Reaction	**"Standard" Reduction Potential**
_____	____ ____ ____ ____
_____	____ ____ ____ ____
_____	____ ____ ____ ____
_____	____ ____ ____ ____
_____	____ ____ ____ ____
_____	____ ____ ____ ____

Optional calculations mean value (each series):

Galvanic Cell	**Mean Value**
_____	_____
_____	_____
_____	_____
_____	

Experiment 21 Galvanic Cells

Galvanic Cell   **Mean Value**

_____   _____

_____   _____

_____   _____

_____   _____

deviation   _____

standard deviation   _____

                        95% confidence level   _____

Effect of Concentration
concentration cell couple used _____

	Calculated	Observed
identical half-cells	_____	_____
potential after 10x dilution	_____	_____
potential after 100x dilution	_____	_____

Show the details of all calculations; use extra sheets if necessary.

QUESTIONS
(Submit your answers on a separate sheet as necessary.)

1. Write the notation and cell reaction for the galvanic cell consisting of your assigned half-cell and the $Pb|Pb^{2+}$ electrode. Draw a diagram showing the direction of electron flow through the circuit. Clearly label the anode, cathode, and salt bridge (or porous membrane).

2. Write the line notation and a cell reaction for each subsequent cell you constructed.

3. When a certain standard half-cell is connected to a standard $Pb|Pb^{2+}$ half-cell, the measured cell potential difference is 0.27 V. The lead half-cell functions as the cathode in this circuit. Compute the cell potential difference when the original half-cell is connected to a standard $Cu|Cu^{2+}$ half-cell.

4. Briefly discuss whether the following experimental "mistakes" would produce a high result, a low result, or have no effect on reported electrode potentials.
    a. The electrolyte in the half-cell undergoing reduction is more dilute than reported.
    b. An aqueous solution of KCl is used as a salt bridge instead of agar gel saturated with KCl.

Experiment 21 Galvanic Cells

Introduction to Qualitative Analysis

The purpose of inorganic qualitative analysis is to identify the cations and anions in a single salt or in a mixture. In modern laboratories, samples containing many ions are routinely analyzed by spectroscopic and chromatographic methods. However, study of the chemical schemes used in Experiments 21-24 is valuable for several reasons: (1) if only a few identifications are needed, the chemical analysis are more convenient and less costly, (2) you will observe some typical reactions of a number of commonly occurring ions, (3) the procedures provide good illustrations of the principles of chemical equilibrium, (4) you will learn some semimicro laboratory techniques.

Cation Analysis (Experiments 22-24)

The first step in the analysis for cations consists of separation of the ions into several groups by taking advantage of some property common to all of the ions in that group. Usually a reaction that produces insoluble compounds is employed. The reagent used for this purpose is called the "group reagent".

Analysis Scheme I shows how group reagents are used to divide some 28 ions into 5 groups. In the diagram, the group reagents are denoted by parentheses, ( ); separations of precipitates from solutions are denoted by branched lines, $\perp$; and the precipitated ions are indicated by double arrows pointed downward, $\Downarrow$.

The separation of group I cations, Ag^+, Hg_2^{2+}, and Pb^{2+}, is achieved by adding dilute HCl to the test solution. The solubility product constants of $AgCl$, Hg_2Cl_2, and $PbCl_2$ are relatively small, and these salts precipitate when a slight excess of chloride ion is added to the solution. The chlorides of all other cations are relatively soluble (see Appendix B Table 1.) Therefore the cations of groups II-IV remain in solution.

The cations of group II are precipitated as sulfides from acidic solution. The precipitating reagent is thioacetamide, TA, which upon heating in water hydrolyzes to give H_2S.

$$CH_3CSNH_2(aq) + 2H_2O \rightleftarrows NH_4^+(aq) + CH_3COO^-(aq) + H_2S\ (g)$$

In aqueous solution, hydrogen sulfide dissociates.

$$H_2S + H_2O \rightleftarrows H_3O^+ + HS^- \text{ with } K_{a1} = \frac{[H_3O^+][HS^-]}{[H_2S]} = 5.7 \times 10^{-8}$$

$$HS^- + H_2O \rightleftarrows H_3O^+ + S^{2-} \text{ with } K_{a2} = \frac{[H_3O^+][S^{2-}]}{[HS^-]} = 1.3 \times 10^{-13}$$

The overall dissociation constant for H_2S, corresponding to the loss of both protons, is
$$K_2 = K_{a1}K_{a2} = \frac{[H_3O^+][HS^-]}{[H_2S]} = 7.4 \times 10^{-21}$$

When an aqueous solution is saturated with H_2S at 1 atm pressure and 25°C, the concentration of dissolved H_2S is 0.10 M. Hence,
$$\frac{[H_3O^+]^2[S^{2-}]}{0.10} = 7.4 \times 10^{-21}$$

Analysis Scheme I
Separation of Cations into Groups

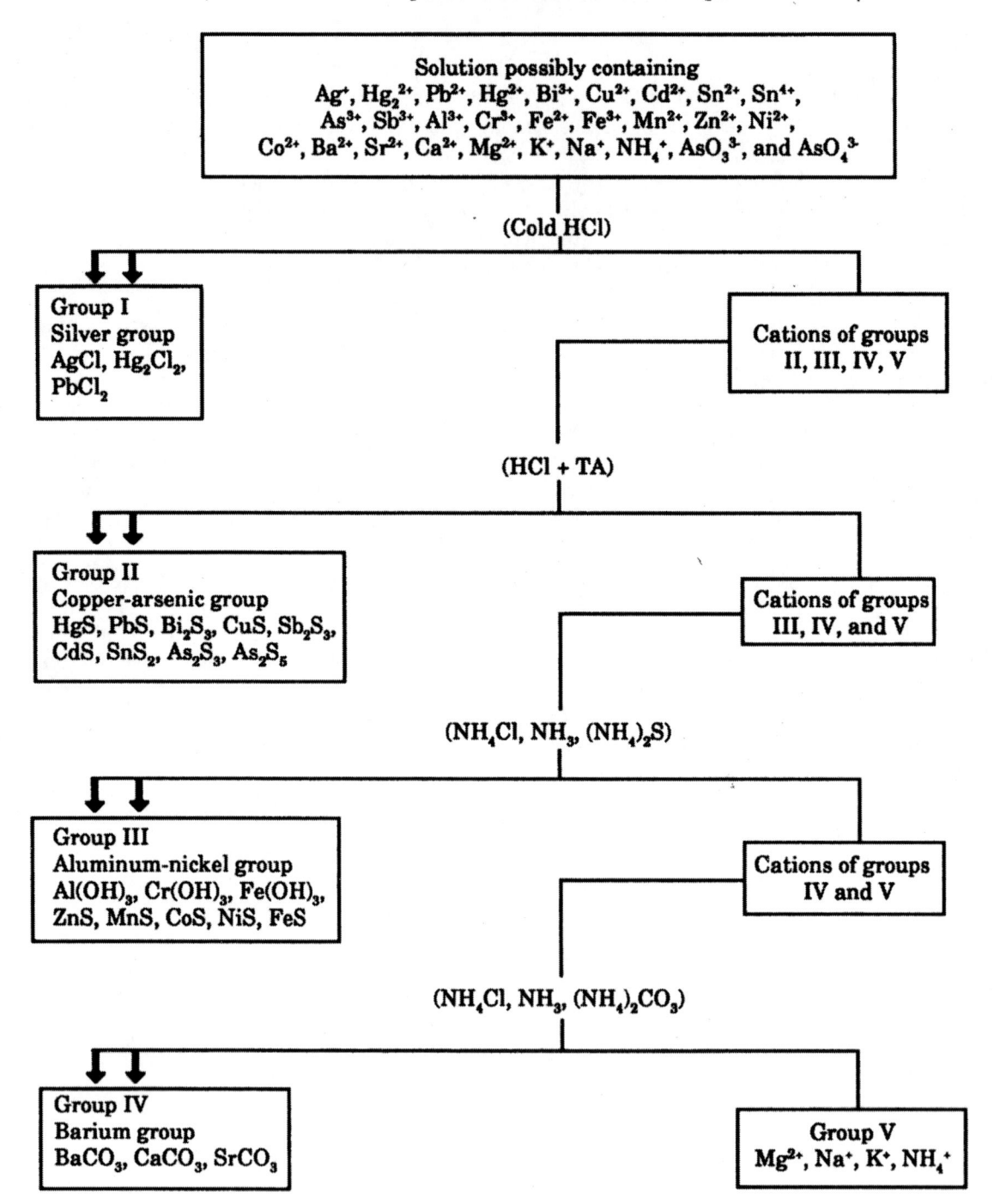

```
┌─────────────────────────────────────────────────────────┐
│ Solution possibly containing                            │
│ Ag⁺, Hg₂²⁺, Pb²⁺, Hg²⁺, Bi³⁺, Cu²⁺, Cd²⁺, Sn²⁺, Sn⁴⁺,   │
│ As³⁺, Sb³⁺, Al³⁺, Cr³⁺, Fe²⁺, Fe³⁺, Mn²⁺, Zn²⁺, Ni²⁺,   │
│ Co²⁺, Ba²⁺, Sr²⁺, Ca²⁺, Mg²⁺, K⁺, Na⁺, NH₄⁺, AsO₃³⁻, and AsO₄³⁻ │
└─────────────────────────────────────────────────────────┘
```

(Cold HCl)

Group I
Silver group
$AgCl$, Hg_2Cl_2, $PbCl_2$

Cations of groups II, III, IV, V

(HCl + TA)

Group II
Copper-arsenic group
HgS, PbS, Bi_2S_3, CuS, Sb_2S_3, CdS, SnS_2, As_2S_3, As_2S_5

Cations of groups III, IV, and V

(NH_4Cl, NH_3, $(NH_4)_2S$)

Group III
Aluminum-nickel group
$Al(OH)_3$, $Cr(OH)_3$, $Fe(OH)_3$, ZnS, MnS, CoS, NiS, FeS

Cations of groups IV and V

(NH_4Cl, NH_3, $(NH_4)_2CO_3$)

Group IV
Barium group
$BaCO_3$, $CaCO_3$, $SrCO_3$

Group V
Mg^{2+}, Na^+, K^+, NH_4^+

Introduction to Qualitative Analysis

$[H_3O^+]^2 [S^{2-}] = 7.4 \times 10^{-22}$ for saturated H$_2$S solution

$$[S^{2-}] = \frac{7.4 \times 10^{-22}}{[H_3O^+]^2}$$

Thus, the sulfide ion concentration of a solution saturated with H$_2$S is inversely proportional to the square of the hydronium ion concentration. Adjusting the hydronium ion concentration produces a wide range of sulfide ion concentrations, with the result that various metal sulfides can be selectively precipitated. For example, the sulfide ion concentration in a solution in which the hydronium ion concentration is approximately 1 M is just sufficient to precipitate CuS ($K_{sp} = 6 \times 10^{-36}$) but not large enough to allow precipitation of NiS ($K_{sp} = 3 \times 10^{-19}$).

It should be noted that the sulfides of group I cations are also insoluble but these cations were removed earlier as chlorides. Because PbCl$_2$ is slightly soluble, it is not possible to completely remove Pb^{2+} along with the other group I cations. A trace of Pb^{2+} is carried over to group II. Therefore, in Analysis Scheme I, lead is encountered in both group I and group II.

The cations of group III are precipitated from basic solution as hydroxides (aluminum subgroup) and sulfides (nickel subgroup). The group reagent is a solution containing NH$_3$, NH$_4^+$, and S^{2-}. An excess of ammonium ion is added to control the hydroxide ion concentration by the equilibrium

$$NH_3 + H_2O \rightleftharpoons NH_4^+ + OH^-$$

Thus, while Al(OH)$_3$ precipitates ($K_{sp} = 1 \times 10^{-33}$), neither Ni(OH)$_2$ ($K_{sp} = 2 \times 10^{-4}$) nor Mg(OH)$_2$ ($K_{sp} = 1 \times 10^{-11}$) will precipitate. However, in the basic solution, the sulfide ion concentration is large enough to precipitate Ni^{2+} and Zn^{2+} as sulfides.

The cations in group IV are precipitated as carbonates. The cations of groups I, II, and III would also form insoluble carbonates had they not been removed previously. In the case of the group IV cations, in order to achieve the necessary high concentration of carbonate ions, the precipitation is carried out at a pH greater than 10.

However if the pH is too high, Mg(OH)$_2$ will also precipitate. Therefore, the solution is maintained at the optimum pH by means of an NH$_3$/NH$_4^+$ buffer.

The group V cations have soluble chlorides, sulfides, hydroxides, (except Mg^{2+} as discussed above), and carbonates. These cations are tested for after all of the other groups have been removed from the original solution. An exception is the ammonium ion which *must* be tested for in a portion of the original solution because ammonium salts and ammonia are used as reagents in several steps of the separations and identifications of other cations.

Anion Analysis (Experiment 25)

In the case of analysis for anions, the group reagents are not used to separate individual ions but instead to indicate which specific ions to test for. For example, silver nitrate is the group reagent for halide ions. If upon adding silver nitrate solution to an unknown, a precipitate is obtained which is insoluble in dilute HNO$_3$, specific tests only for chloride, bromide or iodide need to be performed. The other anions which have silver salts soluble in water or in acid are absent and do not have to be tested.

Qualitative Analysis Scheme Employed in Experiments 22-25

A complete system of inorganic qualitative analysis would include methods for the detection of all known cations and anions. While such an undertaking is beyond the scope of this course, an understanding of the methods and underlying principles can be achieved by studying the identification of a few representative cations and anions. The experiments in this manual, therefore are restricted to the identification of the following ions:

cations: Ag^+, Hg_2^{2+}, Pb^{2+}, Cu^{2+}, Sb^{3+}, Al^{3+}, Ni^{2+}, Ca^{2+}, Ba^{2+}, Mg^{2+}, Na^+, K^+, NH_4^+

anions: Cl^-, Br^-, I^-, NO_3^-, SO_4^{2-}, CO_3^{2-}, PO_4^{3-}, $C_2H_3O_2^-$.

TECHNIQUES OF QUALITATIVE ANALYSIS

Experimental work in qualitative analysis requires careful observations. Before coming to the laboratory, study the experiment to be performed. A knowledge of why a series of steps is performed in a given order will help you to interpret the observed results. You will thus save considerable time and avoid much confusion.

The operations of qualitative analysis are usually performed on a semimicro scale. That is, the volumes of solutions are of the order of 1 to 4 mL, and except for acids and bases, concentrations of solutions are around 0.1 *M* or less. Consequently, it is necessary to work with scrupulously clean apparatus and to avoid contaminating any of the reagents or test solutions. Clean all apparatus with detergent solution, rinse thoroughly with tap water, and finally rinse with distilled water. Distilled water must be used in making up all solutions and in all dilutions of test solutions.

To be successful in the laboratory, you must use accurate techniques of measuring out reagents. Since the quantities of materials are small, you can often adjust the concentrations of solutions by using the proper ratio of *drops* of solution. The droppers in the laboratory kit and in some of the reagent bottles deliver about 1 mL in 20 drops. Capillary droppers which you construct yourself require a larger number of drops to deliver 1 mL. The delivery capacities of these droppers should be determined precisely by counting the number of drops needed to fill a graduated cylinder to a certain volume. Once the capacity of an individual dropper has been established, it should be marked so that it can be easily identified.

Many reagents on the side shelf are equipped with droppers, and these should be used *at the side shelf.* If you need a small quantity of some reagent from a bottle which does not have its own dropper, transfer some of the reagent to a clean beaker or test tube. Then use your own dropper to take from the small sample, and discard what you do not use. *Never put your own dropper into a reagent bottle, and never return unused reagent to the bottle.* Always rinse your dropper with distilled water immediately after using it and then place it on a clean towel.

If you need a solid reagent, tap a few crystals from the reagent bottle into a clean dry beaker. Any excess reagent must be discarded. *Never put a spatula or other object into any reagent bottle.*

When adding one solution to another, always mix them thoroughly by using either a stirring rod or a snapping, shaking motion of the fingers, or by drawing up a little liquid in a dropper and squirting it back into the test tube.

Precipitation Techniques

Initially, solutions should be clear - that is, transparent (not necessarily colorless). If the solution has a cloudiness or turbidity and does not clear up on standing for a few minutes, it should be filtered or centrifuged.

When mixing two reagents that give a precipitate, stir vigorously to encourage formation of the precipitate. Some precipitates are gelatinous and difficult to recognize. In such cases, it is helpful to compare the solution being tested with another sample of the original clear solution to which no reagent has been added.

A precipitate is separated from the solution in which it is formed by filtration or by centrifuging. In the centrifuge, the precipitate settles to the bottom of the centrifuge tube because of centrifugal force on the rapidly whirling solution. When a tube containing a precipitate is put in the centrifuge, a similar tube containing an equal weight of water or solution must be placed in the opposite side of the device as a counterbalance. If the machine is operated when unbalanced, it will vibrate excessively and ultimately its bearings will be ruined. Also, the precipitate will not settle readily in a vibrating centrifuge. If the centrifuge does not operate smoothly, turn off the switch and investigate the cause.

Let the centrifuge slow down gradually from high speeds. Do not attempt to slow it by using your hands, for this will produce unwanted vibrations that disperse the precipitate.

After centrifuging, the precipitate will be firmly packed on the bottom of the centrifuge tube. The clear liquid portion is called the *supernate* or the *centrifugate*. It can be poured off from the densely packed precipitate, a process called decanting; then it is called the *decantate*.

To test for complete precipitation, add another drop of the precipitating reagent to the supernate after centrifuging, while it is still in the centrifuge tube. If cloudiness appears, add more reagent, centrifuge again, and then test again for completeness of precipitation.

If the precipitate is of low density and does not pack firmly in the bottom of the tube, the supernate can be removed by inserting the capillary dropper into the solution and drawing the liquid up gradually so as not to unsettle the precipitate.

Washing precipitates. After decanting the supernate from the precipitate, wash the solid by adding a small amount of water or a specified wash solution. Then shake or stir the mixture, centrifuge, and decant again. The washing step may be repeated several times. Whenever you change the quantity of liquid centrifuged in the tube containing the precipitate, you must change the quantity of liquid in the counterbalancing tube accordingly.

Water Bath Heating

A water bath may be constructed by setting a 250-mL beaker three quarters full of water on a wire gauze over a ring attached to a ring stand. The ring should be no more than one inch above the top of the Bunsen burner barrel. A perforated fiber board is fitted over the beaker and tubes containing mixtures to be heated are inserted into the holes in the fiber board. (One tube may contain distilled water so that a supply of hot water is always available.) If the ring supporting the bath is placed too high above the burner, the fiber board will char and the ring screw will become too hot to touch. Once the water in the bath reaches the boiling point, the burner flame should be adjusted so that boiling is just barely maintained.

Adjusting Acidity

Always stir thoroughly when changing the acidity of a solution. Test periodically by transferring a drop of solution from the stirring rod to moistened litmus paper on a clean watch glass. If the desired pH is acidic, use blue litmus to observe the first change from blue to red. Conversely, when the solution is to be adjusted from an acidic to a basic range, use red litmus top see the change to blue.

Flame Tests

Certain cations in salt vapors impart characteristic colors to a burner flame by emission of visible light. Some examples are sodium (yellow), potassium (violet), barium (yellow-green), calcium (orange-red), copper (blue-green), antimony (bluish-white), and lead (pale blue). Since salts containing ammonium or chromate ions cause flashes in the Bunsen flame, these cations are usually removed before applying a flame test.

The tests are performed with a nichrome or chromel flame test wire mounted in a glass rod. The free end of the wire should be fashioned into a small loop to contain the material being tested. HCl is added to form chloride salts, which are generally more readily volatilized than other salts such as nitrates or sulfates. It is always advisable to compare flame test colors with those from known chloride solutions.

Cleaning the flame test wire before use. Alternately dip the length of the wire into a tube filled with *12 M* HCl and heat to redness in a very hot Bunsen flame to volatilize and remove impurities. Repeat until the flame shows no characteristic cation colors. If the wire is very contaminated and the HCl becomes discolored, replace the acid with a fresh batch. Never put a cleaned wire on the lab bench or in your hand. Store it vertically in a clean test tube or in the HCl.

Testing a *solution.* Adjust the burner flame to remove the inner cone. Dip the loop of the wire into a few drops of solution that has been acidified with one drop of 12 M HCl and bring the loop upward into the lower edge of the flame, so that any color produced seems to be imparted to the entire flame. Repeat several times if necessary. If the solution is very dilute, a larger volume may be tested by feeding a drop of the solution into the loop with a dropper.

Testing a solid. Moisten the solid with 12 M HCl to form a paste. Take up a small amount of the paste in the cleaned wire loop and proceed with the flame test.

Recording Data

All observations and notes on analysis procedures should be recorded directly into a hard-cover bound laboratory notebook. The date should be written in the margin at the beginning of the record for that day. Use the tabular form shown on the report sheets to record your observations of the results obtained for each group reagent or specific test. Colors of all precipitates and solutions are significant information and should always be recorded. An example of a notebook record for Experiment 20 is shown below.

FURTHER READING

MSDS information for the extensive list of qualitative reagents may be found in Appendix E of the Instructor's edition.

Introduction to Qualitative Analysis

SUGGESTED FORM OF NOTEBOOK RECORD FOR QUALITATIVE ANALYSIS

Experiment 22
CATION GROUP I

ANALYSIS SCHEME
Copy Analysis Scheme II

GROUP I KNOWN

Substance	Reagent	Observation	Net Ionic Equations
1. Group I test sol'n (Ag^+, Hg_2^{2+}, Pb^{2+})	6 M HCl	white ppt	$Ag^+ + Cl^- \rightarrow AgCl(s)$ $Hg_2^{2+} + 2Cl^- \rightarrow Hg_2Cl_2(s)$ $Pb^{2+} + 2Cl^- \rightarrow PbCl_2(s)$
2. ppt from 1	hot water	white residue	$PbCl_2(s) \rightarrow Pb^{2+} + 2Cl^-$
3. residue from 2	15 M NH_3	gray-black ppt.	$AgCl(s) + 2NH_3 \rightarrow Ag(NH_3)_2^+ + Cl^-$ $Hg_2Cl_2(s) + NH_3 \rightarrow \ldots\ldots$

GROUP I UNKNOWN

Substance	Reagent	Observation	Inference or Conclusion
Unknown solution		clear and colorless	colored cations absent
1. 5 drops of unknown	6 M HCl	white ppt.	group I present
2. ppt. from 1	hot water	white residue	Ag^+ and/or Hg_2^{2+} present
3. residue from 2	15 M NH_3	gray-black ppt.	Hg_2^{2+} present
4. decantate from 3	16 M HNO_3	no ppt.	Ag^+ absent
5. decantate from 2	K_2CrO_4	no ppt.	Pb^{2+} absent

FINAL CONCLUSION - Hg_2^{2+} present

Introduction to Qualitative Analysis

EQUIPMENT FOR QUALITATIVE ANALYSIS

The following list of equipment is suggested for use in the experiments on qualitative analysis.

Qualitative Analysis Kit

flame test wire, 8-cm	1 iron ring, 3-in.
1 beaker, 150-mL	1 metric rule
1 beaker, 250-mL	1 nichrome triangle
1 clamp, Bunsen	1 pinch clamp
1 clamp special	1 spatula
2 clamp fasteners	1 test-tube brush
1 crucible	1 test-tube holder
1 crucible cover	1 test-tube rack
1 pair of crucible tongs	12 test tubes, 12 x 75-mm
1 evaporating dish	1 thermometer
1 Erlenmeyer flask, 250-mL	5 test tubes, 18 x 150-mm
1 short stem funnel, 65-mm	1 test tube, 25 x 200-mm
1 glass square (cobalt)	3 watch glasses, 25-mm, 65-mm, 100-mm
1 glass cutter	
1 graduated cylinder, 10-mL	1 wing top
2 medicine dropper bulbs	2 wire gauzes
1 perforated fiber board	1 tube of blue litmus paper
wash bottle	1 tube of red litmus paper

Pipets and Stirring Rods

In addition to the equipment provided in the qualitative analysis kit, you will need some items that you must construct yourself. Review the material on glassworking in the *Laboratory Equipment and Techniques* section. Then prepare several stirring rods from a 4-mm outside diameter glass rod cut into 8-cm and 12 cm lengths. Fire-polish both ends of each rod.

Make two capillary pipets by heating the center of 20 cm length of 7-mm outside diameter glass tubing in the Bunsen flame without using a wing top.

REAGENTS FOR QUALITATIVE ANALYSIS

The reagents for qualitative analysis, listed on the following page, are found on a side shelf in the laboratory. After use, they must be returned to the shelf and placed exactly in their original positions. In addition to maintaining neatness, this procedure minimizes the chance that the reagents will become contaminated.

SAFETY AND DISPOSAL

Refer to the MSDS information available online when working with the special reagents for qualitative analysis. MSDS qualitative analysis reagent information (laboratory exercises 22-25) is available online or from your instructor in Appendix E of the Instructor's Edition. Disposal for these compounds and their derived salts should be in accordance with local, state and federal regulations. Disposal for inorganic chemicals should be into a labeled laboratory waste container. Disposal for unreacted acids should be into a labeled laboratory waste jar for acids. Disposal for unreacted bases should be into a labeled laboratory waste jar for bases. Disposal for organic hydrocarbons should be into labeled laboratory waste jars for hydrocarbons.

Introduction to Qualitative Analysis

Special Reagents for Qualitative Analysis

acetic acid, 5 M
acetic acid, 2 M
aluminon reagent, 0.1%
aluminum chloride, 0.2 M
ammonia, 15 M
ammonia, 6 M
ammonia, 5 M
ammonia, 2 M
ammonium acetate, 6 M
ammonium acetate, 0.2 M
ammonium carbonate, 6 M
ammonium chloride, 6 M
ammonium molybdate reagent, 0.5 M
ammonium monohydrogen phosphate, 0.2 M
ammonium oxalate, 0.2 M
ammonium sulfate, 0.2 M
ammonium sulfide reagent, 10%
anion group I test solution
    (mixture of SO_4^{2-}, PO_4^{3-}, CO_3^{2-})
anion group II test solution
    (mixture of Cl^-, Br^-, I^-)
anion group III test solution
    (mixture of NO_3^-, $C_2H_3O_2^-$)
antimony trichloride, 0.2 M
barium hydroxide, saturated
barium nitrate, 0.2 M
calcium chloride, 0.2 M
cation group I test solution
    (mixture of Ag^+, Hg_2^{2+}, Pb^{2+})
cation groups II-V test solution
    (mixture of Cu^{2+}, Sb^{3+}, Al^{3+}, Ni^{2+}, Ba^{2+}, Ca^{2+}, Mg^{2+}, Na^+, K^+, NH_4^+)
chlorine water, saturated
copper(II) chloride, 0.2 M
dimethylglyoxime, 1%

dichloromethane
ethanol, 95%
hydrochloric acid, 12 M
hydrochloric acid, 6 M
hydrochloric acid, 3 M
hydrochloric acid, 1 M
iron(II) sulfate (solid)
lead acetate, 0.5 M
magnesium chloride, 0.2 M
magnesium reagent, 0.01%
mercury(I) nitrate, 0.05 M
nickel chloride, 0.2 M
nitric acid, 16 M
nitric acid, 6 M
nitric acid, 3 M
potassium chloride, 0.2 M
potassium chromate (solid)
potassium chromate, 0.2 M
potassium hexacyanoironate(II), 0.2 M
potassium iodide (solid)
potassium peroxydisulfate (solid)
silver nitrate, 0.2 M
sodium acetate (solid)
sodium acetate, 6 M
sodium bromide (solid)
sodium carbonate (solid)
sodium chloride (solid)
sodium chloride, 0.2 M
sodium hexanitritocobaltate(III) reagent
sodium hydroxide, 8 M
sodium nitrate (solid)
sodium phosphate (solid)
sodium sulfate (solid)
sulfuric acid, 18 M
sulfuric acid, 2 M
thioacetamide, 1 M

246 Name _____ Lab Section _____ Date_____

Introduction to Qualitative Analysis

QUESTIONS

(Submit your answers on a separate sheet as necessary.)

1. For each of the following cation pair mixtures, select the group reagent that will separate the cations and write a net ionic equation for the reaction that occurs:
    a. Ag^+, Cu^{2+}

    b. Cu^{2+}, Cr^{3+}

    c. Hg_2^{2+}, Hg^{2+}

    d. Ba^{2+}, K^+

    e. Al^{3+}, Ca^{2+}

2. Calculate the $[S^{2-}]$ in a saturated solution of H_2S in which the $[H_3O^+]$ is 0.10 M. At 25°C the concentration of H_2S in a saturated solution is 0.10 M.

3. An aqueous solution is 0.0001 M in each of the following cations:

    Ag^+, Sb^{3+}, Cd^{2+}, and Zn^{2+}

    Show by means of appropriate calculations which metal sulfides will precipitate if the solution is maintained about 1×10^{-24} M in sulfide ion.

4. What minimum concentration of NH_4^+ is required in a solution which is 0.10 M in NH_3 and 0.010 M in Ni^{2+}, if the precipitation of $Ni(OH)_2$ is to be avoided?

5. A certain capillary dropper requires 32 drops to deliver 1.0 mL. What is the pH of the solution which results when this dropper is used to add 8 drops of 0.020 M NaOH to 2.0 mL of water?

6. Using the dropper described in question 5, how many drops of 1.0 M NaOH must be added to 2.0 mL of 0.050 M acetic acid to obtain a buffer solution having a pH of 5.00?

Introduction to Qualitative Analysis

Experiment 22
Qualitative Analysis for Cation Group I

OBJECTIVE

To illustrate the use of a group reagent in the separation and identification of the cations in cation group I (Ag^+, Hg_2^{2+}, and Pb^{2+}); to identify the group I cations present in an unknown solution.

EQUIPMENT

See the qualitative analysis kit described in the *Introduction to Qualitative Analysis* section.

REAGENTS

Reagents listed in the *Introduction to Qualitative Analysis* section including the cation group I test solution and unknown group I cation solution.

SAFETY AND DISPOSAL

Refer to the MSDS information available online and disposal recommendations when working with materials found in the qualitative analysis kit described in the *Introduction to Qualitative Analysis* section.

INTRODUCTION

As shown in Analysis Scheme I listed in the *Introduction to Qualitative Analysis* section, when HCl, the group reagent for cation group I, is added to a solution containing all the common cations, only $AgCl$, Hg_2Cl_2, and $PbCl_2$ precipitate. All other cations remain in solution. The flow scheme for the internal analysis of cation group I is shown in Analysis Scheme II. The insoluble chlorides are separated from one another by dissolving $PbCl_2$ in hot water followed by dissolving AgCl in concentrated ammonia solution.

PROCEDURE

Wear your safety goggles throughout the experiment. Before starting the experiment, review the qualitative analysis techniques section in *Introduction to Qualitative Analysis*.

Place 5 drops of cation group I test solution in a 12 x 75-mm labeled test tube (10 to 12 drops when analyzing the unknowns. See Note 22-1.) Add 2 or 3 drops of 6 M HCl, stir, and centrifuge for one or two minutes. To test for complete precipitation, add another drop of HCl to the supernate and observe whether more precipitate forms. If so, cool the solution, add another drop of 6 M HCl, centrifuge, and again test for completeness of precipitation. If no further precipitation occurs, decant the supernate into a clean labeled test tube. (Note 22-2).

Add 1 mL of water to the precipitate, stir thoroughly to form a suspension, and heat the mixture in a boiling water bath for five minutes. Stir the contents of the tube occasionally during the heating to help dissolve the $PbCl_2$. Also place a tube containing distilled water in the bath in order to have hot water available for the washing steps that follow. Centrifuge the hot chloride mixture. Pour the hot supernate into a clean labeled test tube and set aside. This solution is to be tested for Pb^{2+} ion.

Wash the remaining precipitate twice with several drops of hot water, mixing, heating, centrifuging, and decanting as before (Note 22-3). Discard the hot water washings and save the precipitate to test for Ag^+ and Hg_2^{2+} ions.

Test for Hg_2^{2+}. To the precipitate add 5 drops of water and 5 drops of 15 M NH_3 solution. Stir thoroughly and centrifuge. A black or very dark gray precipitate of Hg (black) plus $HgNH_2Cl$ (white) confirms the presence of mercury (I). Ammonium ions and chloride ions are byproducts in this reaction.

Decant the ammoniacal solution into a clean test tube. This solution should contain silver as the complex ion $[Ag(NH_3)_2]^+$.

Test for Ag^+. To the ammoniacal solution, add 16 M HNO_3 *dropwise* with stirring until the mixture tests acidic to blue litmus. (Test for acidity by touching the stirring rod to a piece of moistened blue litmus paper. Do *not* dip the litmus paper into the solution.) A white precipitate of AgCl that appears upon acidification confirms the presence of silver.

Test for Pb^{2+}. To the decantate from the initial hot water treatment, add 2 drops of K_2CrO_4 solution. A yellow precipitate of $PbCrO_4$ confirms the presence of lead.

Unknown. Obtain an unknown solution from your instructor. Carry out the procedure for the cation group I analysis and identify the ion(s) present.

ANALYSIS SCHEME II

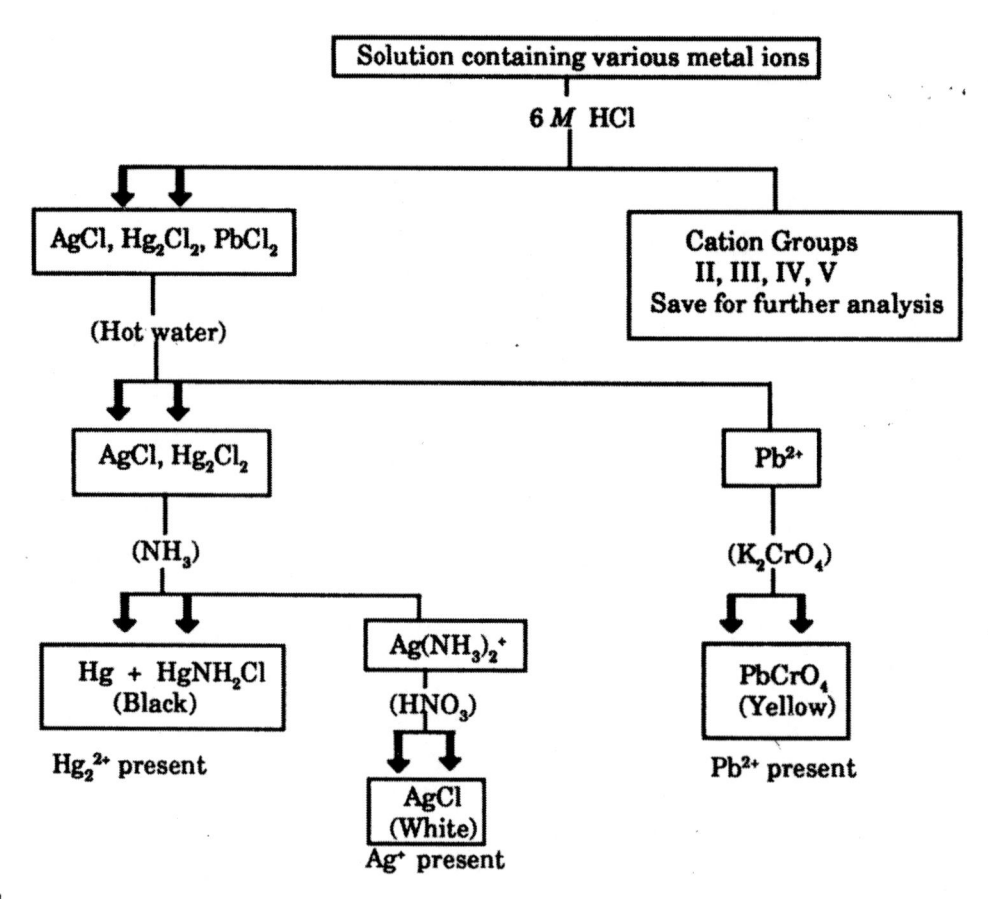

```
                Solution containing various metal ions
                              |
                           6 M HCl
              ┌───────────────┴───────────────┐
     AgCl, Hg₂Cl₂, PbCl₂              Cation Groups
              |                       II, III, IV, V
         (Hot water)                Save for further analysis
              |
      ┌───────┴───────┐
   AgCl, Hg₂Cl₂       Pb²⁺
        |              |
       (NH₃)        (K₂CrO₄)
        |              |
   ┌────┴────┐      PbCrO₄
Hg + HgNH₂Cl  Ag(NH₃)₂⁺   (Yellow)
  (Black)       |
              (HNO₃)     Pb²⁺ present
Hg₂²⁺ present   |
              AgCl
             (White)
            Ag⁺ present
```

NOTES

22-1. If no precipitate is observed immediately upon the addition of HCl, vigorously rub the inner walls of the test tube with a stirring rod for a few minutes. If this does not produce a precipitate, cations of group I may be considered to be absent.

22-2. If the solution contains only group I cations, as in this experiment, the decantate may be discarded. However, if this decantate may contain the cations of groups II-V, it must be labeled and saved for further analysis. It is then advisable to add the following step to the procedure: wash the chloride precipitate with 5 drops of cold water, centrifuge, and add this washing to the decantate containing groups II-V.

22-3. If the precipitate is not thoroughly washed free of lead ions, the Pb^{2+} may interfere with the test for Hg_2^{2+} by reacting with ammonia in the next step to form the white, insoluble $Pb(OH)_2$.

Experiment 22 Qualitative Analysis for Cation Group I

Name_____ Lab Section_____ Date_____

Prelaboratory Assignment: Experiment 22
Qualitative Analysis of Cation Group I

1. What cations are found in Group I?

2. What reagent(s) are used to separate Group I from Groups II--V?

3. In what groups are mercury ions found? Why are they found in different groups?

4. How are lead compounds separated from other group I compounds?

Experiment 22 Qualitative Analysis for Cation Group I

5. If 6 M nitric acid is added to a solution of silver ions and lead ions, will a precipitate form. If so, what will be its composition?

6. If 6 M sulfuric acid is added to a solution that is 1 M each in silver ions and lead ions, will a precipitate form. If so, what will be the composition of the precipitate?

Name _____ Lab Section_____ Date_____ 251

REPORT ON EXPERIMENT 22
Qualitative Analysis for Cation Group I

DATA AND RESULTS

Group I Known

Substance	Reagent	Observation	Net Ionic Equations

Group I Unknown

Substance	Reagent	Observation	Net Ionic Equations

Cations Present _____

Experiment 22 Qualitative Analysis for Cation Group I

QUESTIONS

(Submit your answers on a separate sheet as necessary.)

1. The solubility product constant for $PbCl_2$ is 1.7×10^{-5}. Calculate the solubility of $PbCl_2$ (in moles per liter) in pure water and in a solution that is $0.1\ M$ in Cl^- ion. Assume that all $PbCl_2$ that dissolves is present as Pb^{2+} and Cl^- ions. Repeat these calculations for AgCl, for which the solubility product constant is 1.8×10^{-10}.
2. Explain, using net ionic equilibrium equations and the LeChatelier principle, why adding nitric acid to an ammoniacal solution containing Ag^+ and Cl^- causes AgCl to precipitate.
3. In the analysis for the group I cations, explain why it would be inadvisable to test for lead by adding K_2CrO_4 to the original unknown solution.
4. For each of the following group I cation unknowns, infer which cation(s) must be absent and which must be present:
    a) an unknown in which the group I chloride precipitate dissolves completely in hot water
    b) an unknown in which the washed residue after the hot water treatment dissolves completely in $15\ M\ NH_3$.

Experiment 22 Qualitative Analysis for Cation Group I

Experiment 23
Qualitative Analysis for Cation Groups II-V

OBJECTIVE
To demonstrate the principles and procedures for separating and identifying cations in groups II-V.

EQUIPMENT
See the qualitative analysis kit described in the *Introduction to Qualitative Analysis* section.

REAGENTS
Reagents listed in the *Introduction to Qualitative Analysis* section including the cation groups II-V test solution

SAFETY AND DISPOSAL
Refer to the MSDS information available online and disposal recommendations when working with materials found in the qualitative analysis kit described in the *Introduction to Qualitative Analysis* section.

INTRODUCTION
This experiment is concerned with the general procedures shown in Analysis Scheme I described in the *Introduction to Qualitative Analysis* section for separating the group II-V cations into individual groups. Emphasis is on the experimental conditions required for selectively precipitating only the ions in a given group rather than on the identification of a large number of individual ions within each group. Consequently, analysis for only some of the cations from each group will be done. For instructions on how to test for other cations, consult the textbooks on qualitative analysis listed under Further Reading in the Introduction to Qualitative Analysis. (The cations in this experiment are colorless in aqueous solution except for Cu^{2+} (blue or green) and Ni^{2+} (green).

Group II cations are precipitated from bulk solutions as sulfides when hydrogen sulfide is added in the presence of about 1 M H_3O^+. In this experiment the bulk solution will contain only Cu^{2+} and Sb^{3+} ions as representative of the two subgroups within cation group II. The antimony ion (representative of the subgroup consisting of antimony, arsenic and tin) forms soluble complex ions in the presence of excess sulfide ion. Thus, antimony sulfide dissolves in ammonium sulfide, while the sulfides of the copper subgroup (copper, mercury (II), lead, cadmium, and bismuth) do not. In this manner the two subgroups, of which antimony and copper are representative, may be separated.

Group III cations are then separated from the bulk solution as hydroxides or sulfides. The coprecipitation of group III sulfides with group II sulfides can be prevented by careful control of the pH of the solution during the group II separation earlier. In this experiment the bulk solution will contain only Al^{3+} and Ni^{2+} ions, which are representative of the two subgroups within group III. The aluminum ion (representative of the subgroup containing aluminum, iron(III), and chromium(III) is precipitated as a hydroxide in a weakly basic solution buffered with NH_3 and NH_4Cl. Then the nickel ion (representative of the subgroup containing nickel, iron(II), zinc, cobalt, and manganese) is precipitated as a sulfide by the addition of $(NH_4)_2S$ (Note 23-1). In this experiment, the procedure used for separating Al^{3+} and Ni^{2+} ions is simplified in the absence of the other group III cations. When the latter are present, it is necessary to consult the more detailed procedures for the separation of group III precipitates described in sources cited under Further Reading in the Introduction to Qualitative Analysis. Here it is more convenient to precipitate $Al(OH)_3$ and then add $(NH_4)_2S$ to the decantate to precipitate NiS.

Group IV cations are separated from solution by precipitation as carbonates at a pH of about 10. Buffering the solution again with NH_3 and NH_4Cl prevents the precipitation of $Mg(OH)_2$, that would occur at a higher pH. Barium ions and calcium ions have been selected as representative of group IV.

The cations of group V (Mg^{2+}, Na^+, K^+, and NH_4^+), which have soluble chlorides, sulfides, and carbonates, remain in solution. Although ammonium ion, NH_4^+, is included in this group, it must always be tested for on a portion of the *original* solution before analysis for the other cations. Why?

PROCEDURE

Wear your safety goggles throughout the experiment. Before starting the experiment, study the qualitative analysis techniques sections in the Introduction to Qualitative Analysis. Refer frequently to Analysis Scheme III shown at the end of this experiment. Associate the various operations you perform with the steps outlined in this flow scheme. Record your observations directly in your notebook in tabular form.

Separation of Group II: Tests for Cu^{2+} and Sb^{3+}

Put 8 drops of cation groups II-V test solution (or in the case of an unknown solution all of the decantate from the precipitation step of group I cations) in an evaporating dish on a wire gauze over a ring mounted on a ring stand. Using a small blue flame without an inner cone, heat very gently under the hood and evaporate the solution to a thick paste but *not* to dryness (Note 23-2). Add 8 drops of 3 M HCl and stir (warming if necessary) until the solution is clear. Transfer this solution to a 12 x 75-mm test tube. Wash the evaporating dish with 10 drops of 3 M HCl and transfer the washing to the same test tube. Add 1 drop of 0.2 M ammonium acetate (Note 23-3) and 20 drops of thioacetamide (TA). Stir and heat the test tube in a boiling water bath for at least 10 minutes **under the hood**. Observe the progressive changes in color that take place (Note 23-4).

Centrifuge, and decant the clear supernate into a clean test tube. This solution contains the cations of groups III-V. Label it and set it aside. Wash the group II precipitate twice with 20-drop portions of 0.2 M ammonium acetate. Discard the washings.

To the test tube containing the precipitate (CuS and Sb_2S_3), add 10 drops of ammonium sulfide reagent to dissolve the Sb_2S_3. Stir the mixture thoroughly. Wash down with ammonium sulfide any precipitate adhering to the sides of the test tube, centrifuge, and decant, saving the decantate to analyze for antimony. Repeat the treatment of the precipitate with a second 10-drop portion of ammonium sulfide reagent. Centrifuge and decant. Combine the two ammonium sulfide decantates, label the test tube, and set it aside.

A remaining black precipitate indicates copper sulfide. Wash it twice with 20-drop portions of a hot water solution containing 5 drops of 0.2 M ammonium acetate. Discard the washings.

Add 15 drops of 3 M HNO_3 to the washed precipitate. Mix thoroughly and boil in a water bath for about 1 minute or until all reaction ceases. The insoluble material that remains is sulfur. Centrifuge and transfer the decantate to a fresh test tube. (If the sulfur, which has a low density, floats on the surface of the liquid, it may be removed with a stirring rod or, alternatively, the liquid may be extracted with a capillary dropping pipet.)

To the clear decantate, add 15 M NH_3 dropwise, stirring after the addition of each drop, until the solution is decidedly alkaline to litmus paper. If a precipitate forms, centrifuge and discard the precipitate. A completely colorless decantate at this point (view the solution vertically above a white background if necessary) indicates that copper is absent and need not be tested for further. If the decantate is deep blue due to the $Cu(NH_3)_4^{2+}$ complex ion, copper is present. Place 5 drops of this blue solution in a test tube. Add 5 M acetic acid until the deep blue color just disappears, and then add 2 drops of 0.2 M potassium hexacyanoironate (II), $K_4[Fe(CN)_6]$. A red precipitate, $Cu_2[Fe(CN)_6]$, further confirms the presence of Cu^{2+}.

To the test tube containing the solution of Sb_2S_3 in ammonium sulfide (mainly antimony as the complex ion $(Sb_2S_3^{3-})$, add 6 M HCl until the solution is slightly acidic to litmus paper. Centrifuge and decant, discarding the decantate. Wash the precipitate with 30 drops of hot water. Then add 10 drops of 12 M HCl and heat in a water bath for about 5 minutes with stirring (Note 23-5). If any precipitate remains, centrifuge and transfer the clear decantate to a clean test tube. Discard the precipitate. To the clear solution add 20 drops of water and 10 drops of TA; stir and heat in the water bath **under the hood** for about 5 minutes. An orange precipitate of Sb_2S_3 indicates the presence of Sb^{3+} ion.

Separation of Group III: Tests for Al^{3+} and Ni^{2+}

Transfer the solution containing groups III-V cations to a clean evaporating dish. **Under the hood** boil gently (avoid splattering) for about 5 minutes to decompose all the TA in the solution and release H_2S gas (Note 23- 6). The hydrogen sulfide may be detected by holding a moistened piece of lead acetate paper (or a filter paper impregnated with a few drops of lead acetate reagent) above the evaporating dish. If the paper darkens due to the formation of black PbS from the reaction between H_2S and Pb^{2+} ion, H_2S gas is still emanating from the solution. Boil until there is no further test with lead acetate paper, but *do not* evaporate to dryness; add distilled water if necessary. Transfer the solution to a test tube. Wash the evaporating dish with 10 to 12 drops of water, transferring the washing to the same test tube. Centrifuge, and discard any precipitate.

To the *clear* solution add 5 drops of 6 M NH_4Cl, *stir* and then add 2 M NH_3 *dropwise with constant stirring* until the solution just turns basic to litmus. A white gelatinous precipitate of $Al(OH)_3$ indicates the presence of Al^{3+}ion. Centrifuge (if you have a precipitate), and transfer the supernate (containing Ni^{2+} ion and cations of groups IV and V) to a clean test tube. Label this tube and set it aside.

To confirm the presence of Al^{3+} ion wash the white gelatinous precipitate twice with 20 to 25-drop portions of water. Dissolve the washed precipitate in a few drops of 3 M HNO_3. Add 1 drop of aluminon reagent and stir. Then add 5 M NH_3 until the solution is just basic to litmus and centrifuge. A cherry red precipitate (due to the absorption of the red aluminon dye on $Al(OH)_3$) confirms the presence of Al^{3+} ion (Note 23-7).

To the solution containing Ni^{2+} and cations of groups IV and V, add 4 drops of $(NH_4)_2S$ reagent and stir. A black precipitate of NiS indicates the presence of Ni^{2+} ion. The colored $(NH_4)_2S$ solution may obscure this precipitate, which may not be visible until the mixture is centrifuged and the liquid decanted off.) Centrifuge for a few minutes, and transfer the supernate (which contains cations of groups IV and V) to a crucible and set it aside.

To confirm the presence of nickel, wash the black precipitate twice with 30-drop portions of water. Drain off the last washing as much as possible. To the washed precipitate, add 6 drops of 12 M HCl and 2 drops of 16 M HNO_3 (this 3: 1 mixture of HCl and HNO_3 is called *aqua regia*). Mix and warm gently in a water bath under the hood. If a precipitate remains, it is sulfur that was formed by the oxidation of sulfide ions. Centrifuge, and transfer the supernate to a clean test tube. Discard the precipitate. To the clear supernate, add 15 drops of water and 4 drops of dimethylglyoxime reagent. Stir, and then add 15 M NH_3 dropwise with stirring until the solution is basic to litmus. A strawberry red precipitate confirms the presence of Ni^{2+} ion (Note 23-8).

Separation of Group IV: Tests for Ba^{2+} and Ca^{2+}

In a crucible supported by a wire triangle, evaporate the decantate containing cation groups IV and V until solids begin to crystallize. Add 1 mL of 16 M HNO_3 by pouring the acid down the sides of the crucible so that most of the solid material is washed down. Transfer the ring stand and crucible to a hood and heat again, cautiously, until the contents are dry. Then heat more strongly until no more fumes of ammonium salts are driven off.

After the crucible has cooled, add 5 drops of 3 M HCl and 10 drops of water. Stir and transfer the solution to a 12 x 75-mm test tube. Rinse the crucible with another 10-drop portion of water and combine the rinsings with the solution in the test tube. If the resulting solution is not clear, centrifuge and discard the precipitate.

Transfer the clear solution to a test tube, add 4 drops of 6 M NH_4Cl, and with stirring make the solution barely alkaline with 2 M NH_3 (Note 23-9). Then add 5 drops of 6 M ammonium carbonate, stir, and let the mixture stand for 5 minutes. If the solution does not become turbid due to the formation of a fine precipitate, vigorously rub the inner walls of the test tube with a stirring rod for a few minutes. (If the solution remains completely clear, group IV can be considered absent, and you should proceed to the group V analysis.) Centrifuge and test for completeness of precipitation by adding an additional drop of $(NH_4)_2CO_3$ to the supernate. Decant the supernate, which contains cations of group V, into a clean labeled test tube, and save it for further analysis. The precipitate consists of barium and calcium carbonates.

Dissolve the group IV carbonate precipitate in 2 drops of 6 M HCl. Make this solution barely alkaline with 2 M NH_3 (ignore any precipitate which may form at this point), and then add 2 M acetic acid dropwise until you have added 1 drop more than is necessary for complete neutralization (Note 23-10). Add 4 drops of water and mix thoroughly. Then add 1 drop of 0.2 M potassium chromate solution and stir; if a fine yellow precipitate forms, barium is present. Centrifuge the mixture. (The amount of $BaCrO_4$ may be very small and not evident until after centrifuging.) In the event of a precipitate, add more chromate solution dropwise to the supernate until precipitation is complete. Decant, and save the decantate for the analysis of calcium.

After washing the $BaCrO_4$ precipitate twice with 5-drop portions of hot water, dissolve it in 4 drops of 6 M HCl. Perform a flame test on this solution and look for a yellow-green flame of very short duration to further confirm the presence of barium.

To the decantate from the barium chromate precipitation (or if Ba^{2+} is absent, the solution of the group IV cations), add 2 M NH_3 dropwise until the solution is alkaline. Then add 5 drops of ammonium oxalate solution. A white precipitate, CaC_2O_4, shows the presence of calcium. Centrifuge, and discard the supernate. Wash the precipitate three times with hot water, dissolve it in a few drops of 6 M HCl, and run a flame on the solution. An orange-red flame confirms the presence of calcium. Run a comparison flame test on a $CaCl_2$ solution from the reagent shelf.

Analysis of Group V: Tests for Mg^{2+}, K^+, Na^+, and NH_4^+:

Treat the decantate from the group IV carbonate precipitation with 2 drops of ammonium sulfate solution and 2 drops of ammonium oxalate solution. Heat to boiling, centrifuge, and decant, discarding any precipitate (Note 23-11). Divide the solution into two portions. Treat one portion with 4 drops of ammonium monohydrogen phosphate solution, $(NH_4)_2HPO_4$. Mix thoroughly and warm gently in a warm water bath. A white crystalline precipitate, $MgNH_4PO_4$, shows the presence of magnesium. Wash the precipitate three times with hot water, and dissolve it in a few drops of 6 M HCl. To this solution add

2 drops of para-nitrobenzene azoresorcinol (magnesium reagent), and then add 8 M sodium hydroxide solution until the solution is alkaline. Formation of a blue flocculent precipitate (which is best viewed by centrifuging and discarding the supernatant liquid) confirms the presence of magnesium (Note 23- 12).

Under the hood evaporate the remaining half of the solution of group V cations to dryness in a crucible, and bake the residue until the evolution of dense white fumes subsides, indicating that all of the ammonium salts have been sublimed. After the crucible has cooled, add 2 drops of 6 M HCl and perform a flame test. A persistent fluffy yellow flame shows the presence of sodium. If a fairly diffuse yellow flame of short duration is observed, the sodium is merely present as an impurity and not as a *bona fide* component of the material being tested. A lavender flame of short duration shows the presence of potassium and the absence of sodium. This lavender flame is obscured by the yellow flame in cases where sodium is present. However, viewing the flame through a blue cobalt glass, which filters out the yellow radiation, permits the potassium flame to be visible as a reddish violet color. Run comparison tests with KCl and NaCl solutions from the reagent shelf.

The test for the ammonium ion must be performed on the original test solution. On the convex side of a watch glass large enough to cover a 150-mL beaker, affix a strip of red litmus paper with a drop of distilled water. Place a few drops of the original test solution in the beaker and, without splashing, add about 5 drops of 8 M NaOH. Swirl the beaker gently to mix and cover it promptly with the watch glass, having the litmus paper on the underside. If the litmus paper turns uniformly blue within 30 seconds, NH_4^+ was present in the solution. For further evidence, remove the watch glass and waft the pungent fumes of generated NH_3 gas cautiously toward your nose. In this test, extreme care must be exercised to avoid contact between the alkaline NaOH solution and the litmus. Also, if a heavy precipitate forms in the beaker, more sodium hydroxide should be added.

An Additional Test for K^+ Ion

In the absence of NH_4^+ ion. To a few drops of the original test solution add an equal amount of 6 M $NaC_2H_3O_2$. If any precipitate forms, centrifuge and discard the precipitate. To the clear solution add a few drops of sodium hexanitritocobaltate(III) reagent. The presence of K^+ is indicated by a yellow precipitate of $K_3[Co(NO_2)_6]$.

In the presence of NH_4^+ ion. To a few drops of the original test solution in an evaporating dish, add enough 8 M NaOH to make it fairly basic. Gently heat the evaporating dish until all the NH_4^+ has been converted to ammonia (*i.e.*, until a moist red litmus paper held over the evaporating dish does not turn blue). Take care not to let the solution dry up; if necessary, add water. Now acidify the solution with 6 M HCl and add a few drops of $Na_3[Co(NO_2)_6]$ reagent.

A yellow precipitate shows the presence of K' ion. The NH_4^+ must be removed first, as described, because it also gives a yellow precipitate with the same reagent.

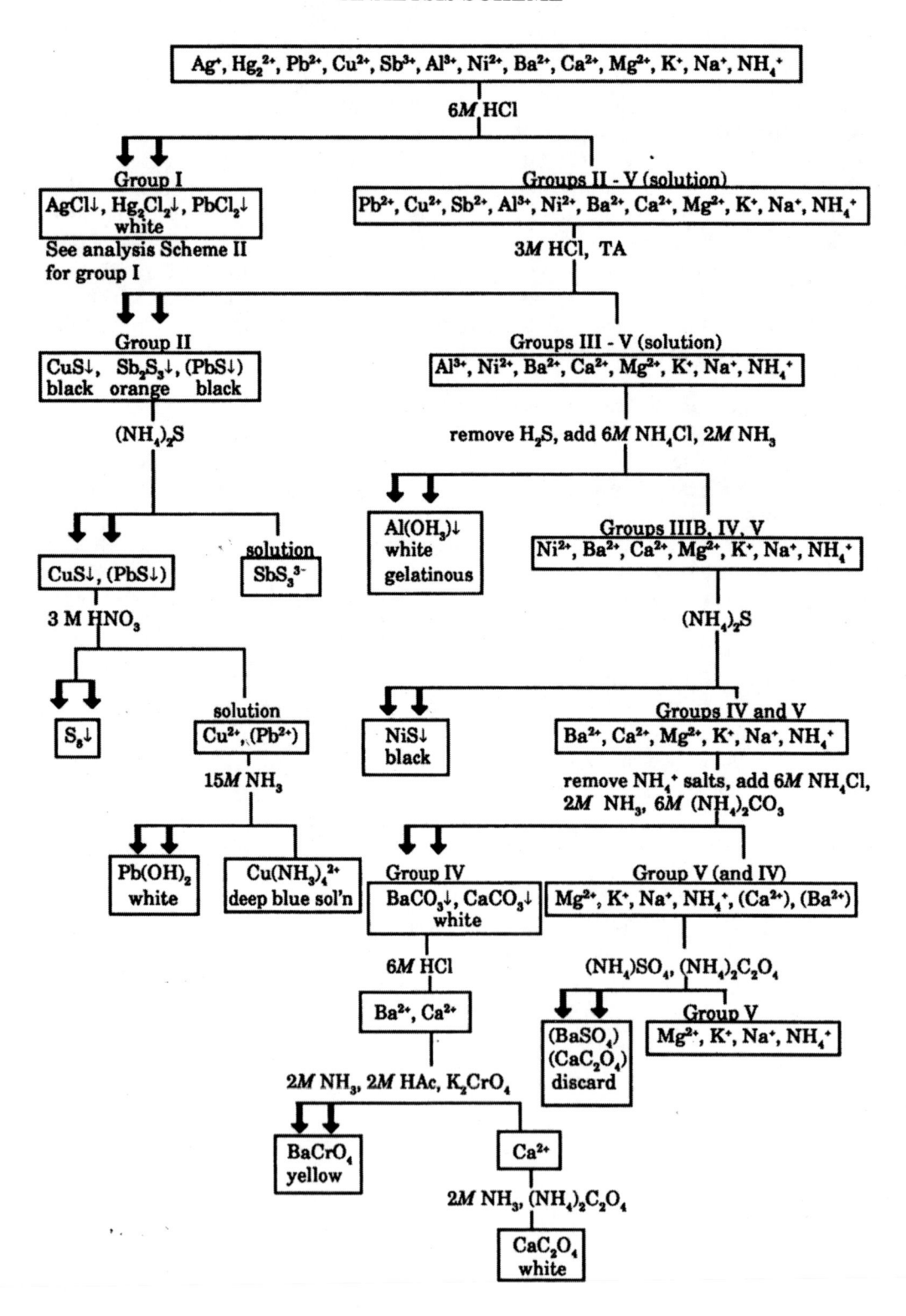

ANALYSIS SCHEME

```
Ag⁺, Hg₂²⁺, Pb²⁺, Cu²⁺, Sb³⁺, Al³⁺, Ni²⁺, Ba²⁺, Ca²⁺, Mg²⁺, K⁺, Na⁺, NH₄⁺
                            │
                         6M HCl
            ┌───────────────┴───────────────┐
         Group I                   Groups II - V (solution)
   AgCl↓, Hg₂Cl₂↓, PbCl₂↓     Pb²⁺, Cu²⁺, Sb²⁺, Al³⁺, Ni²⁺, Ba²⁺, Ca²⁺, Mg²⁺, K⁺, Na⁺, NH₄⁺
          white                            │
   See analysis Scheme II              3M HCl, TA
       for group I        ┌────────────────┴───────────────────┐
                       Group II                    Groups III - V (solution)
                 CuS↓, Sb₂S₃↓, (PbS↓)         Al³⁺, Ni²⁺, Ba²⁺, Ca²⁺, Mg²⁺, K⁺, Na⁺, NH₄⁺
                 black  orange  black                    │
                       │                    remove H₂S, add 6M NH₄Cl, 2M NH₃
                   (NH₄)₂S               ┌───────────────┴──────────────┐
              ┌─────┴─────┐            Al(OH)₃↓              Groups IIIB, IV, V
         CuS↓, (PbS↓)   solution         white        Ni²⁺, Ba²⁺, Ca²⁺, Mg²⁺, K⁺, Na⁺, NH₄⁺
              │         SbS₃³⁻        gelatinous                │
          3 M HNO₃                                           (NH₄)₂S
          ┌───┴────┐                              ┌──────────────┴─────────────┐
         S₈↓   solution                         NiS↓                 Groups IV and V
              Cu²⁺, (Pb²⁺)                     black         Ba²⁺, Ca²⁺, Mg²⁺, K⁺, Na⁺, NH₄⁺
                │                                                      │
              15M NH₃                                    remove NH₄⁺ salts, add 6M NH₄Cl,
          ┌────┴──────┐                                   2M NH₃, 6M (NH₄)₂CO₃
      Pb(OH)₂    Cu(NH₃)₄²⁺                  ┌──────────────────┴──────────────────┐
       white    deep blue sol'n          Group IV                Group V (and IV)
                                       BaCO₃↓, CaCO₃↓      Mg²⁺, K⁺, Na⁺, NH₄⁺, (Ca²⁺), (Ba²⁺)
                                           white                       │
                                             │              (NH₄)SO₄, (NH₄)₂C₂O₄
                                          6M HCl              ┌────────┴────────┐
                                         Ba²⁺, Ca²⁺        (BaSO₄)          Group V
                                             │            (CaC₂O₄)     Mg²⁺, K⁺, Na⁺, NH₄⁺
                                  2M NH₃, 2M HAc, K₂CrO₄    discard
                                      ┌──────┴─────┐
                                   BaCrO₄        Ca²⁺
                                   yellow          │
                                           2M NH₃, (NH₄)₂C₂O₄
                                                   │
                                                CaC₂O₄
                                                 white
```

Experiment 23 Cation Groups II-V

NOTES

23-1. In this experiment, the procedure used for separating Al^{3+} and Ni^{2+} ions is simplified in the absence of the other group III cations. When the latter are present, it is necessary to consult more detailed procedures for the separation of group III precipitates. Here it is convenient to precipitate $Al(OH)_3$ and then add $(NH_4)_2S$ to the decantate to precipitate NiS.

23-2. Evaporating to dryness should be avoided because some of the cation salt residue might be lost by volatilization or splattering. The purpose of evaporating to a paste is to boil off any acid present and to reduce the solution volume. Only if this is accomplished can the desired pH be obtained by adding the prescribed amount of 3 M HCl in the next step.

23-3. Ammonium acetate prevents the formation of a colloidal dispersion of very fine solid particles that do not settle out.

23-4. If the test tube is more than three-fourths full, excessive frothing may cause the mixture to overflow into the water bath. If necessary, use two test tubes to heat the mixture and combine the portions later.

23-5. The precipitate formed when 6 M HCl is added to the ammonium sulfide solution contains sulfur and, if antimony is present, Sb_2S_3 as well. The sulfur is formed from the action of the HCl on ammonium polysulfide, $(NH_4)_2S_2$, an impurity present in the $(NH_4)_2S$ solution. When this precipitate is treated with 12 M HCl, the Sb_2S_3 dissolves to form the complex ion $[SbCl_4^-]$ and H_2S. The yellow or white sulfur portion of the precipitate does not dissolve in HCl and is discarded. The supernate containing the $[SbCl_4^-]$ ion is treated later with TA to reprecipitate the pure orange Sb_2S_3 free of sulfur contamination.

23-6. If the H_2S is not removed, some NiS may precipitate prematurely along with $Al(OH)_3$ when the solution is made basic with ammonia.

23-7. The solution should appear nearly colorless after centrifuging with the red dye absorbed to the aluminum hydroxide precipitate at the bottom of the tube.

23-8. Dimethylglyoxime, $(CH_3)_2C_2(NOH)_2$, reacts with Ni^{2+} in weakly ammoniacal solution to form a red chelate compound in which the nickel ion is bound to four nitrogen atoms.
$$2\,(CH_3)_2C_2(NOH)_2 + Ni^{2+} + 2NH_3 \rightarrow NH_4^+\,Ni(C_4H_7N_2O_2)_2$$
Since the red compound is soluble in acid, the solution must be basic for precipitation to occur. However, if too much NH_3 is added, the formation of the violet blue complex ion $[Ni(NH_3)_6^{2+}]$ is favored, and the red precipitate may not appear. Therefore, the addition of the 15 M NH_3 must be performed with care.

23-9. The solution must be very carefully buffered. If it is too acidic, the group IV carbonates will not precipitate; if it is too basic, $Mg(OH)_2$ will precipitate along with $BaCO_3$ and $CaCO_3$.

23-10. The pH must be very carefully adjusted at this point to produce a chromate ion concentration that is just sufficient to precipitate the very insoluble $BaCrO_4$ and repress the precipitation of the more soluble $CaCrO_4$. Adding acid reduces the CrO_4^{2-} concentration according to the chromate-dichromate equilibrium in aqueous solution.

$$2CrO_4^{2-} + 2H_3O^+ \rightleftharpoons Cr_2O_7^{2-} + 3H_2O$$

The dichromate salts are relatively soluble. Also, if the solution is basic the white $Ca(OH)_2$ may precipitate.

23-11. The group IV carbonate precipitation does not remove all of the Ba^{2+} and Ca^{2+} ions from solution. Therefore, the decantate is treated at this stage with $(NH_4)_2SO_4$ and $(NH_4)_2C_2O_4$, to remove trace amounts of barium and calcium. These group IV cations would interfere with the test for Mg^{2+} by forming insoluble phosphates in the analysis of group V.

23-12. The solution must be alkaline for the formation of $Mg(OH)_2$. The magnesium reagent dye, which is very selective for magnesium, then is adsorbed on the magnesium hydroxide.

Experiment 23 Cation Groups II-V

Name _____ Lab Section _____ Date _____ 261

Prelaboratory Assignment: Experiment 23
Qualitative Analysis of Cation Group II- V

1. Which cations belong to the following:
    a. Group II

    b. Group III

    c. Group IV

    d. Group V

2. What reagents are used to separate Group II from Groups III, IV, and V?

3. What reagents are used to separate Group III from Groups IV, and V?

4. What reagents are used to separate Group IV from Group V?

Experiment 23 Cation Groups II-V

5. In testing for potassium ion, why is it necessary or desirable to know if ammonium ion is present?

6. In performing a flame test for the potassium ion, why is it a good idea to observe the flame through a blue cobalt glass?

Name _____ Lab Section _____ Date _____   263

REPORT ON EXPERIMENT 23
Qualitative Analysis for Cation Groups II--V

DATA AND RESULTS

Substance	Reagent	Observation	Net Ionic Equations

Experiment 23 Cation Groups II-V

DATA AND RESULTS (Cation Groups II-V)

Substance	Reagent	Observation	Net Ionic Equations

Experiment 23 Cation Groups II-V

QUESTIONS

(Submit your answers on a separate sheet as necessary.)

1. Assuming a dipositive cation, M^{2+}, what would be the maximum K_{sp} of its sulfide, MS, if it is to be precipitated from a 0.01 M solution of M^{2+} by adding H_2S at a pH of 1?

2. Compare the K_{sp} calculated above with those of tin(II) sulfide and the α form of nickel(II) sulfide. Would these sulfides precipitate under the conditions given above? Explain.

3. Nitric acid oxidizes sulfide ion to sulfur. Explain on the basis of net ionic equilibrium equations and LeChatelier's principle why CuS dissolves more readily in HNO_3 than in HCl.

4. By means of appropriate calculations, determine whether the nickel precipitate obtained at pH 8 was $Ni(OH)_2$ or NiS.

5. Explain in terms of Le Chatelier's principle why ammonium ion must be added to an ammonia solution in order to prevent the precipitation of $Mg(OH)_2$ along with the carbonates of the group IV cations.

6. List at least two advantages of using ammonium salts as reagents in various steps of the qualitative analysis scheme rather than the more readily available sodium salts.

COMMENTS

Experiment 23 Cation Groups II-V

Experiment 24
Analysis of a General Cation Unknown

OBJECTIVE

To apply the principles of qualitative analysis in devising a scheme for identifying several cations in a mixture.

EQUIPMENT

See the qualitative analysis kit described in the *Introduction to Qualitative Analysis* section.

REAGENTS

Reagents listed in the *Introduction to Qualitative Analysis* section.

SAFETY AND DISPOSAL

Refer to the MSDS information available online and disposal recommendations when working with materials found in the qualitative analysis kit described in the *Introduction to Qualitative Analysis* section.

INTRODUCTION

In this experiment you will analyze an unknown containing at least 5 cations. Unless otherwise instructed, test for these 12 ions:

Ag^+, Hg^{2+}, Pb^{2+}, Cu^{2+}, Sb^{3+}, Ni^{2+}, Al^{3+}, Ba^{2+}, Ca^{2+}, Mg^{2+}, K^+, NH_4^+

The general procedures of the previous two experiments should be followed. However, if the schemes are thoroughly understood some of the procedures may be shortened considerably. Also, it is advisable to first perform flame tests and to test a portion of the unknown for NH_4^+. Careful observations at this stage will be quite useful. For example, if in testing for NH_4^+ a precipitate is formed, additional information can be acquired by adding an excess of NaOH to see whether the precipitate dissolves. If the precipitate dissolves completely or if no precipitate forms in the first place, several possible ions are immediately eliminated.

PROCEDURE

Wear your safety goggles. Secure an unknown solution from your instructor. Following the procedures of Experiments 22 and 23, identify the cations present. Compare tests for specific ions with tests on authentic samples from the side shelf. Record all tests and observations in your notes, and complete the report sheet.

COMMENTS

Prelaboratory Assignment: Experiment 24
Qualitative Analysis of a General Cation Unknown

1. In what cation groups are each of the following ions classified:

   Ag^+, Hg^{2+}, Pb^{2+}, Cu^{2+}, Sb^{3+}, Ni^{2+}, Al^{3+}, Ba^{2+}, Ca^{2+}, Mg^{2+}, K^+, NH_4^+

2. Why is it is advisable to first perform flame tests on the general cation unknown?

Experiment 24 Analysis of a General Cation Unknown

3. Why is it advisable to test a portion of the unknown for NH_4^+?

4. Devise and diagram a flow scheme with the minimum number of steps for the separation and identification of only these cations: Pb^{2+}, Cu^{2+}, Sb^{3+}, Ni^{2+}.

Name _____ Lab Section _____ Date _____   271

REPORT ON EXPERIMENT 24
Analysis of a General Cation Unknown

DATA AND RESULTS
(Submit your data and results on a separate sheet if necessary.)

Record the results of your preliminary flame tests and of the test for NH_4^+

Substance	Reagent	Observation	Inference or Conclusion

Cations present _____

Experiment 24 Analysis of a General Cation Unknown

QUESTIONS

(Submit your answers on a separate sheet as necessary.)

1. Select a *single* reagent that can separate each of the following pairs and write net ionic equations for any reactions that occur.

    a. Sb_2S_3, NiS

    b. $AgCl$, $Al(OH)_3$

    c. $Al(OH)_3$, $Mg(OH)_2$

2. When a blue unknown solution was treated with 6 M HCl, a white precipitate (A) formed. The supernate (A^1) was saved for further analysis. The precipitate (A) was treated with hot water leaving a residue (B). This residue (B) dissolved completely in 15 M NH_3 to give a solution (B^1), in which a white precipitate formed upon acidification. When the supernate (A^1) was evaporated and treated with HCl and TA, a dark precipitate (C) appeared. The supernate (C^1) was set aside. The precipitate (C) was insoluble in ammonium sulfide but dissolved in hot nitric acid. Addition of concentrated ammonia to the resulting solution gave a solution (D^1) with a deep blue color. No precipitation was observed when the supernate (C^1) was boiled and treated with NH_4Cl, NH_3, and $(NH_4)_2S$.

    a. List the cations that are definitely present.

    b. List the cations that are definitely absent.

    c. Identify by formula the precipitates (B) and (C).

    d. Identify by formula the group I - V cationic species in solutions (B^1) and (D^1).

3. Suppose you are to analyze a solution that may possibly contain one or more of the five cations:

    Hg_2^{2+}, Al^{3+}, Ni^{2+}, Ca^{2+}, Mg^{2+}

    Devise and diagram a flow scheme with the minimum number of steps for the separation and identification of only these cations.

Experiment 25 Qualitative Analysis for Common Anions

Experiment 25
Qualitative Analysis for Common Anions

OBJECTIVE
To demonstrate the methods and techniques of testing for and identifying common anions.

EQUIPMENT
See the qualitative analysis kit described in the *Introduction to Qualitative Analysis* section.

REAGENTS
Reagents listed in the *Introduction to Qualitative Analysis* section including anion group I, anion group II, and anion group III test solutions, and other reagents on the qualitative analysis side shelf.

SAFETY AND DISPOSAL
Refer to the MSDS information available online and disposal recommendations when working with materials found in the qualitative analysis kit described in the *Introduction to Qualitative Analysis* section.

INTRODUCTION
In this experiment, the anions that will be investigated are:

$$SO_4^{2-}, CO_3^{2-}, CrO_4^{2-}, PO_4^{3-}, Cl^-, Br^-, I^-, NO_3^-, C_2H_3O_2^-$$

Generally, it is possible to analyze for anions without a separation scheme. Instead a series of tests is performed that narrows the possibilities to one or two anions, which then may be tested for specifically. One procedure, for example, is to (1) perform preliminary tests with concentrated sulfuric acid, (2) attempt to prepare insoluble barium and silver salts as precipitates, and (3) test for the specific anions that are indicated by the two preceding operations. The sulfuric acid tests may be omitted when only a single anion is known to be present. However, this test is an essential aid in analyzing an anion mixture.

Anions may be classified into three groups according to their solubility properties. (Only a few representative members of each group are listed below.)

Group 1. Those forming insoluble barium salts (sulfate group)
$$SO_4^{2-}, CO_3^{2-}, PO_4^{3-}$$

Group II. Those forming soluble barium salts but insoluble silver salts (halide group)
$$Cl^-, Br^-, I^-$$

Group III. Those forming soluble barium and silver salts (nitrate group)
$$NO_3^-, C_2H_3O_2^-$$

The purpose of the group reagents for anions is not to separate individual ions but to indicate which tests for specific anions should be performed. For example, suppose the anion group reagent $Ba(NO_3)_2$ is added to a neutral solution and no precipitate appears. It may be concluded that group I anions are absent and need not be tested for specifically. If, on the other hand, a precipitate *does* appear, specific tests for only the sulfate group are indicated.

In reactions of anions with Ba^{2+} or with Ag^+ ions, care must be taken to maintain the correct pH.

For example, the insoluble barium salts of group I anions dissolve readily in acid except for $BaSO_4$. Therefore, the treatment with $Ba(NO_3)_2$ must be made in neutral or mildly alkaline solution if the precipitation of the carbonate or phosphate salts is to be observed. On the other hand, the silver salts of all anions listed except nitrate, acetate, and sulfate are insoluble in water. Except for the halide salts, however, they dissolve readily in 3 M HNO_3. Hence, in the specific tests for the halide group the test solutions are first acidified to avoid precipitation of silver salts not in the halide group.

In the preliminary tests for anions, concentrated H_2SO_4 acts (1) as a strong acid and (2) as an oxidizing agent. As a strong acid, H_2SO_4 can liberate the acids of other anions (except sulfate) from their salts. In some cases, this liberated acid may then be oxidized by the H_2SO_4 to produce a substance that may be detected by its color or odor. Heating drives off gaseous products and also enhances any oxidation process. Characteristic reactions of a few common anions with 18 M H_2SO_4 are described in Table 25-1.

Anion	Reaction with Cold H_2SO_4	Reaction with Hot H_2SO_4
$C_2H_3O_2^-$	no reaction apparent	odor of vinegar (acetic acid)
SO_4^{2-}	no reaction	no reaction
PO_4^{3-}	no reaction	no reaction
NO_3^-	no reaction	liberation of brown gas and HNO_3, which can be detected by placing a few shreds of Cu in the mouth of the test tube; formation of green $Cu(NO_3)_2$ and brown fumes
Cl^-	effervescence, colorless	sharp odor of HCl gas fumes in moist air
Br^-	effervescence, brown fumes	rapid evolution of Br_2 when heated
I^-	effervescence, solid turns dark instantly	odor of ~S, violet fumes I_2
CO_3^{2-}	effervescence, colorless and odorless gas (test for CO_2)	no further reaction

Table 25-1. Reactions of Anions with Sulfuric Acid

PROCEDURE

Wear safety goggles throughout the experiment. Before starting the experiment, review the qualitative analysis techniques section in Introduction to Qualitative Analysis. On your report sheet write your observations and net ionic equations for all the aqueous solution reactions you observe in each test. For tests in concentrated H_2SO_4 write ordinary chemical equations (Note 25-1). Also, present your observations in tabular form similar to Table 25-1. When your instructor has approved your work, you will be issued an unknown solid that you must identify according to the procedure outlined in Experiment 25.

Preliminary Tests with H_2SO_4

Place 0.01-g samples (about a small spatula) of each of the following salts into separate, clean, dry 12x75-mm test tubes.

sodium sulfate	potassium iodide	sodium nitrate
sodium carbonate	sodium phosphate	sodium acetate
sodium chloride	sodium bromide	

Moisten each sample with a few drops of concentrated H_2SO_4. Observe any immediate reaction, then warm the mixture gently. *Do not boil.* Observe any color change, odor, or evolution of gas. Record the results on your report sheet in the form of a table. Also, write a balanced equation for the reaction of each salt with sulfuric acid.

Caution: avoid breathing deeply any unknown gas, and be careful not to point the heated tube toward yourself or your neighbor. To test for odor, hold the test tube vertically near your nose and fan the vapors toward you, sniffing cautiously.

Treatment of Anion Solutions with $Ba(NO_3)_2$

Place 3 to 4 drops of a solution of each of the nine anions in separate 12x75-mm test tubes. Make each solution alkaline with a drop of 6 M NH_3 and then add 2 drops of 0.2 M $Ba(NO_3)_2$ solution to each. Centrifuge the precipitates, discard the supernates, and wash the precipitates twice with cold water. Attempt to dissolve each of the precipitates in 4 to 5 drops of 6 M HCl. Tabulate your results, noting the colors of all precipitates formed. Write ionic equations for the formation of the precipitates. Write ionic equations for those cases in which the barium salts are dissolved by dilute HCl.

Treatment of Anion Solutions with $AgNO_3$

Place 3 to 4 drops of a solution of each of the anions, separately, in nine test tubes. Add 2 drops of 0.2M $AgNO_3$ solution to each. Centrifuge the precipitates formed. Also, as part of your table, write ionic equations for the formation of the precipitates and note which ones are *not* dissolved by dilute HNO_3, (Note 25-2).

Specific Tests for Group I Anions (Sulfate Group)

Obtain 3 mL of anion group I test solution from the side shelf and perform the following tests:

Sulfate. To 5 drops of the group I solution, add 3 drops of dilute HCl (1 M) and 2 drops of barium nitrate solution. A white precipitate, insoluble in HCl, proves that SO_4^{2-} is present.

Phosphate. To 5 drops of the group I solution, add 5 drops of 16 M HNO_3 and 15 to 20 drops ammonium molybdate solution. Stir, heat nearly to boiling, and allow to stand for a few minutes. Formation of a finely divided canary yellow precipitate, $(NH_4)_3PO_4\cdot12MoO_4$, indicates the presence of PO_4^{3-} (Note 25-3).

Carbonate. To 2 mL of group I test solution, add a few drops of 12 M HCl. Test the escaping gas for CO_2 by holding a drop of barium hydroxide solution, suspended from the tip of a medicine dropper, a short distance down into the mouth of the test tube. The "clouding" of the drop because of the formation of a white precipitate of barium carbonate proves the presence of CO_3^{2-}. Write the net ionic equations for the formation of CO_2 and for the reaction between CO_2 and aqueous barium hydroxide.

Specific Tests for Group II Anions (Halide Group)

Obtain 3 mL of anion group II test solution from the side shelf and perform the following test:

Chloride, Bromide, and Iodide. To 5 drops of the group II test solution, add a few drops of 3M nitric acid, warm gently, and then add 2 drops of 0.2 M $AgNO_3$ Centrifuge and decant the supernate. Add 5 drops of 6 M NH_3 to the precipitate and stir. (If all of the precipitate dissolves at this point, only chloride ion is present). Centrifuge the ammoniacal solution and decant the supernate into a fresh tube. Acidify the supernate with 6 M HNO_3. Formation of a white precipitate indicates Cl^-,

To a separate portion of the group II test solution (3 to 5 drops) add 10 drops of fresh chlorine water followed by 5 drops of dichloromethane. Shake well. A brown color in the dichloromethane layer indicates Br^-: pink or violet indicates I^-. (Note 25-4).

Experiment 25 Qualitative Analysis for Common Anions

If both bromide and iodide ions are present, the violet color of iodine in the dichloromethane may mask the brown. Hence, the following procedure may be used to establish the presence of both ions. Place another portion of the group II test solution (10 to 15 drops) in a casserole, add 15 drops of 5 M acetic acid and 30 drops of water, and mix thoroughly. Then add a small quantity of solid potassium peroxydisulfate, $K_2S_2O_8$ and heat. A brown color, due to the liberation of iodine, proves the presence of iodide. (Note 25-5).

Boil the solution in the casserole under the hood until all of the iodine is removed, replenishing the liquid with water. When the iodine is completely removed, transfer the solution to a test tube and add 5 drops of chlorine water and 5 drops of dichloromethane. Shake well. Brown coloration of the dichloromethane layer proves the presence of Br^-.

Specific Tests for Group III Anions (Nitrate Group)

Obtain 3 mL of anion group III test solution from the side shelf and make the following tests:

Nitrate. To 5 drops of the group III test solution, add 8 to 10 drops of 18 M H_2SO_4; then cool the test tube under the water tap. Now incline the tube at an angle of about 30° and add 1 drop of freshly prepared saturated iron(II) sulfate solution very slowly so that it runs down the side of the tube and forms a layer which floats on the sulfuric acid solution. Bring the tube slowly into an upright position. The formation of a brown ring, $FeSO_4 \cdot NO$, at the junction of the two solutions proves that NO_3^- is present, (Note 25-6).

Acetate. To 5 drops of the group III solution, add 1 mL of concentrated H_2SO_4, and warm the tube slowly to nearly boiling. Cool the tube and add 5 drops of ethyl alcohol; then heat the tube in a boiling water bath for 1 minute. A sweet, fruity odor (ethyl acetate) indicates $C_2H_3O_2^-$. If a solid unknown is being tested the following additional test may be employed. Heat strongly a small amount of solid in a test tube and observe whether there is charring of the material. Charring is characteristic of salts of organic acids and is indicated by darkening of the solid accompanied by acrid fumes.

NOTES

25-1. In concentrated sulfuric acid, most of the H_2SO_4 exists in molecular form.

25-2. Although Ag_2CO_3 is yellow, it tends to darken rapidly in aqueous solution due to its decomposition to CO_2 and black Ag_2O.

25-3. Both phosphates and arsenates react with ammonium molybdate, $(NH_4)_2MoO_4$, to form insoluble yellow precipitates, ammonium phosphomolybdate, $(NH_4)_3PO_4 \cdot 12MoO_4$, or ammonium arsenomolybdate, $(NH_4)_3AsO_4 \cdot 12MoO_4$. As a consequence, arsenates must be removed before the confirmatory test for phosphates can be made. However, in this experiment, for simplicity, arsenate has been omitted.

25-4. Chlorine water, an aqueous solution of Cl_2 gas, oxidizes Br^- ions to molecular Br_2 and I^- ions to I_2. In aqueous solution both Br_2 and I_2 appear brown because of a secondary reaction between iodine and water, and the halogens are not distinguishable in the medium. The dichloromethane, which is immiscible with water, plays no part in the reaction; it serves merely to indicate the presence of bromine and iodine by the characteristic brown and violet colors of these halogens in this solvent. Thorough mixing enhances the extraction of the Br_2 and I_2 from the aqueous phase to the CH_2Cl_2 phase.

25-5. In acetic acid solution the $K_2S_2O_8$ selectively oxidizes I^-, but will not oxidize Br^- ion.

25-6. Iron(II) sulfate solutions are unstable due to the slow oxidation of Fe^{2+} by oxygen in the air. Prepare a fresh saturated solution by dissolving the solid in a few mL of water.
In the test for NO_3^- ion, the Fe^{2+} ions reduce NO_3^- to NO in the presence of sulfuric acid. The excess ferrous ions that were not oxidized to ferric then react with the NO to form the brown complex $[Fe(H_2O)_5NO]^{2+}$. The solution must be cooled because this complex decomposes at higher temperatures.

Name _____ Lab Section _____ Date _____

Prelaboratory Assignment: Experiments 25
Qualitative Analysis of Common Anions

1. List at least two representative members for each of the three groups of anions.

2. What is the purpose of the group reagents for anions?

3. Why must treatment of anions with $Ba(NO_3)_2$ be made in neutral or mildly alkaline solution?

4. Why do silver salts dissolve readily in 3 M HNO_3? Why is fresh chlorine water critical to testing group II anions?

Experiment 25 Qualitative Analysis for Common Anions

5. What two tests are available for acetate ion?

6. Devise with the minimum number of steps for the identification of only anions generated from a solution containing sodium carbonate, sodium phosphate, and sodium acetate.

Name _____ Lab Section _____ Date _____   279

REPORT ON EXPERIMENT 25
Qualitative Analysis for Common Anions

DATA AND RESULTS

I. Write equations for all of the reactions you observed with concentrated H_2SO_4.

Anion	Cold H_2SO_4	Hot H_2SO_4

II. Treatment with $AgNO_3$

Anion	Observation	Net Ionic Equation	Observation upon *Addition of HNO_3*	Net Ionic *Equation*

Experiment 25 Qualitative Analysis for Common Anions

III. Treatment with Ba(NO$_3$)$_2$

Anion	Observation	Net Ionic Equation	Observation upon Addition of *HNO$_3$*	Net Ionic *Equation*

IV. Specific Tests for Anions

Anion	Reagents	Net Ionic Equation
SO$_4^{2-}$	_____	_____
PO$_4^{3-}$	_____	_____
CO$_3^{2-}$	_____	_____
CrO$_4^{2-}$	_____	_____
Cl$^-$	_____	_____
Br$^-$	_____	_____
I$^-$	_____	_____
NO$_3^-$	_____	_____
C$_2$H$_3$O$_2^-$	_____	_____

Experiment 25 Qualitative Analysis for Common Anions

Appendix A
Names and Formulas of Common Ions

Ionic Charge	Name	Symbol
1+	ammonium	NH_4^+
	hydrogen	H^+
	potassium	K^+
	silver	Ag^+
	sodium	Na^+
2+	barium	Ba^{2+}
	calcium	Ca^{2+}
	chromium	Cr^{2+}
	copper(II)	Cu^{2+}
	iron(II)	Fe^{2+}
	lead(II)	Pb^{2+}
	magnesium	Mg^{2+}
	manganese (II)	Mn^{2+}
	mercury(I)	Hg_2^{2+}
	mercury(II)	Hg^{2+}
	tin(II)	Sn^{2+}
	strontium	Sr^{2+}
	zinc	Zn^{2+}
3+	aluminum	Al^{3+}
	chromium(III)	Cr^{3+}
	iron(III)	Fe^{3+}

Ionic Charge	Name	Symbol
1-	acetate	$C_2H_3O_2^-$
	chlorate	ClO_3^-
	chloride	Cl^-
	chlorite	ClO_2^-
	cyanide	CN^-
	fluoride	F^-
	hydride	H^-
	hydrogen carbonate (bicarbonate)	HCO_3^-
	hydroxide	OH^-
	hypochlorite	ClO^-
	iodate	IO_3^-
	iodide	I^-
	nitrate	NO_3^-
	nitrite	NO_2^-
	perchlorate	ClO_4^-
	permanganate	MnO_4^-
2-	carbonate	CO_3^{2-}
	chromate	CrO_4^{2-}
	dichromate	$Cr_2O_7^{2-}$
	oxalate	$C_2O_4^{2-}$
	oxide	O^{2-}
	sulfate	SO_4^{2-}
	sulfide	S^{2-}
	sulfite	SO_3^{2-}
3-	arsenate	AsO_4^{3-}
	arsenite	AsO_3^{3-}
	phosphate	PO_4^{3-}

Appendices A, B, C, D

Appendix B
Tables of Data

1. General Solubility Rules for Common Salts and Hydroxides
2. Saturated Vapor Pressure of Water from 0° to 100°C
3. Relative Density of Water
4. Prefixes for Fractions and Multiples of SI Units
5. Equilibrium Constants for Acidic and Basic Dissociation at 25°C
6. Solubility Products at 25°C
7. Concentration of Side-Shelf Acid and Base Solutions
8. Standard Reduction Potentials at 25°C
9. Common Buffer Solutions
10. Color Changes, pH Intervals, and H_3O^+ Concentration Intervals of Some Important Indicators

Appendices A, B, C, D

TABLE 1
GENERAL SOLUBILITY RULES FOR COMMON SALTS AND HYDROXIDES

Na^+, K^+, NH_4^+	All salts of sodium, potassium, and ammonium are soluble except several uncommon ones such as $Na_4Sb_2O_7$, $K_3Co(NO_2)_6$, K_2PtCl_6, and $(NH_4)_2PtCl_6$.
$C_2H_3O_2^-$	All acetates are soluble, ($AgC_2H_3O_2$ only moderately.)
Cl^-	All chlorides are soluble except $AgCl$, Hg_2Cl_2, and $PbCl_2$. ($PbCl_2$ is slightly soluble in cold water, moderately soluble in hot water.)
NO_3^-	All nitrates are soluble.
SO_4^{2-}	All sulfates are soluble except $BaSO_4$ and $PbSO_4$ ($CaSO_4$, Hg_2SO_4, and Ag_2SO_4 are slightly soluble; the corresponding bisulfates are more soluble.)
Ag^+	All silver salts are insoluble except $AgNO_3$ and $AgClO_4$. ($AgC_2H_3O_2$ and Ag_2SO_4. are only moderately soluble.)
CO_3^{2-} and PO_4^{3-}	All carbonates and phosphates are insoluble except those of Na^+, K^+, and NH_4^+. (Many acid phosphates are soluble, *e.g.*, $Mg(H_2PO_4)_2$ and $Ca(H_2PO_4)_2$.)
OH^-	All hydroxides are insoluble except $NaOH$, KOH, and $Ba(OH)_2$, ($Ca(OH)_2$ is slightly soluble.)
S^{2-}	All sulfides are insoluble except those of Na^+, K^+, and NH_4^+ and those of the alkaline earths, Mg^{2+}, Ca^{2+}, Sr^{2+}, and Ba^{2+}. (Sulfides of Al^{3+} and Cr^{3+} hydrolyze and precipitate the corresponding hydroxides.)

Appendices A, B, C, D

TABLE 2
SATURATED VAPOR PRESSURE OF WATER FROM 0° TO 100

Temperature (°C)	Pressure (torr)	Temperature (°C)	Pressure (torr)	Temperature (°C)	Pressure (torr)
0°	4.6	24°	22.4	40°	55.3
5°	6.5	25°	23.8	50°	92.5
10°	9.2	26°	25.2	60°	149.3
15°	12.8	27°	26.7	70°	233.7
16°	13.6	28°	28.3	80°	355.1
17°	14.5	29°	30.0	90°	525.8
18°	15.5	30°	31.8	98°	707.3
19°	16.5	31°	33.7	98.5°	720.2
20°	17.5	32°	35.7	99.0°	733.2
21°	18.6	33°	37.7	99.5°	746.5
22°	19.8	34°	39.9	100.0°	760.0
23°	21.0	35°	42.2		

TABLE 3
RELATIVE DENSITY OF WATER

Temperature (°C)	Density (g/mL)	Temperature (°C)	Density (g/mL)	Temperature (°C)	Density (g/mL)
0	0.99987	22	780	44	066
1	993	23	756	45	0.99025
2	997	24	732	46	0.98982
3	999	25	0.99707	47	940
4	1.00000	26	681	48	896
5	0.99999	27	654	49	852
6	997	28	626	50	0.98807
7	993	29	597	51	762
8	988	30	0.99567	52	715
9	981	31	537	53	669
10	0.99973	32	505	54	621
11	963	33	473	55	0.98573
12	952	34	440	60	0.98324
13	940	35	0.99406	65	0.98059
14	927	36	371	70	0.97781
15	0.99913	37	336	75	0.97489
16	897	38	299	80	0.97183
17	880	39	262	85	0.96865
18	862	40	0.99224	90	0.96534
19	843	41	186	95	0.96192
20	0.99823	42	147	100	0.95838
21	802	43	107		

Source: Adapted from Smithsonian Tables, compiled from various authors. Note: The mass of 1 cm^3 of water at 4°C is taken at unity. The values given are numerically equal to the absolute density in g/ml.

TABLE 4
PREFIXES FOR FRACTIONS AND MULTIPLES OF SI UNITS

Fraction	Prefix	Symbol	Multiple	Prefix	Symbol
10^{-1}	deci	d	10	deka	da
10^{-2}	centi	c	10^2	hecto	h
10^{-3}	milli	m	10^3	kilo	k
10^{-6}	micro	μ	10^6	mega	M
10^{-9}	nano	n	10^9	giga	G
10^{-12}	pico	p	10^{12}	tera	T
10^{-15}	femto	f	10^{15}	peta	P
10^{-18}	atto	a	10^{18}	exa	E
10^{-21}	zepto	z	10^{21}	zetta	Z
10^{-24}	yocto	y	10^{24}	yotta	Y

TABLE 5
EQUILIBRIUM CONSTANTS FOR ACIDIC AND BASIC DISSOCIATION AT 25°C

Compound	Dissociation Reaction		K_a	pK_a
Water	$H_2O \rightleftharpoons H^+ + OH^-$		1.00×10^{-14}	14.00
Weak acids				
acetic	$HC_2H_3O_2 \rightleftharpoons H^+ + C_2H_3O_2^-$		1.76×10^{-5}	4.75
boric	$H_3BO_3 \rightleftharpoons H^+ + H_2BO_3^-$		6.0×10^{-10}	9.22
carbonic ($CO_2 + H_2O$)	$CO_2 + H_2O \rightleftharpoons H^+ + HCO_3^-$	K_1:	4.4×10^{-7}	6.35
	$HCO_3^- \rightleftharpoons H^+ + CO_3^{2-}$	K_2:	4.7×10^{-11}	10.33
chromic	$H_2CrO_4 \rightleftharpoons H^+ + HCrO_4^-$	K_1:	2×10^{-1}	0.7
	$HCrO_4^- \rightleftharpoons H^+ + CrO_4^{2-}$	K_2:	3.2×10^{-7}	6.50
formic	$HCHO_2 \rightleftharpoons H^+ + CHO_2^-$		2.1×10^{-4}	3.68
hydrofluoric	$HF \rightleftharpoons H^+ + F^-$		6.9×10^{-4}	3.16
hydrogen peroxide	$H_2O_2 \rightleftharpoons H^+ + HO_2^-$		2.4×10^{-12}	11.62
hydrogen sulfate ion	$HSO_4^- \rightleftharpoons H^+ + SO_4^{2-}$	K_2:	1.2×10^{-2}	1.92
hydrogen sulfide	$H_2S \rightleftharpoons H^+ + HS^-$	K_1:	1.0×10^{-7}	7.00
	$HS^- \rightleftharpoons H^+ + S^{2-}$	K_2:	1.3×10^{-13}	12.89
lactic	$HC_3H_5O_3 \rightleftharpoons H^+ + C_3H_5O_3^-$		1.4×10^{-4}	3.85
nitrous	$HNO_2 \rightleftharpoons H^+ + NO_2^-$		4.5×10^{-4}	3.50
oxalic	$H_2C_2O_4 \rightleftharpoons H^+ + HC_2O_4^-$	K_1:	3.8×10^{-2}	1.42
	$HC_2O_4^- \rightleftharpoons H^+ + C_2O_4^{2-}$	K_2:	5.0×10^{-5}	4.30
phosphoric	$H_3PO_4 \rightleftharpoons H^+ + H_2PO_4^-$	K_1:	7.1×10^{-3}	2.15
	$H_2PO_4^- \rightleftharpoons H^+ + HPO_4^{2-}$	K_2:	6.3×10^{-8}	7.20
	$HPO_4^{2-} \rightleftharpoons H^+ + PO_4^{3-}$	K_3:	4.4×10^{-13}	12.36
phosphorus	$H_2HPO_3 \rightleftharpoons H^+ + HHPO_3^-$	K_1:	1.6×10^{-2}	1.80

Compound	Dissociation Reaction		K_b	pK_b
Weak bases				
ammonia	$NH_3 + H_2O \rightleftharpoons NH_4^+ + OH^-$		1.8×10^{-5}	4.75
barium hydroxide	$Ba(OH)_2 \rightleftharpoons BaOH^+ + OH^-$	K_1:	strong	
	$BaOH^+ \rightleftharpoons Ba^{2+} + OH^-$	K_2:	1.4×10^{-1}	0.85
calcium hydroxide	$Ca(OH)_2 \rightleftharpoons CaOH^+ + OH^-$	K_1:	strong	
	$CaOH^+ \rightleftharpoons Ca^{2+} + OH^-$	K_2:	3.5×10^{-2}	1.5
methylamine	$CH_3NH_2 + HOH \rightleftharpoons CH_3NH_3^+ + OH^-$		5×10^{-4}	3.3
pyridine	$C_5H_5N + H_2O \rightleftharpoons C_5H_5NH^+ + OH^-$		1.4×10^{-9}	8.84

Appendices A, B, C, D

TABLE 6
SOLUBILITY PRODUCTS AT 25°C

Solid	K_{sp}	Solid	K_{sp}
AgBr	5×10^{-13}	Hg_2Br_2	7×10^{-18}
Ag_2CO_3	8×10^{-12}	Hg_2Cl_2	7×10^{-23}
$AgC_2H_3O_2$	3×10^{-3}	Hg_2I_2	5×10^{-29}
AgCN	1×10^{-16} *	$Hg_2(OH)_2$	2×10^{-24}
AgCl	2×10^{-10}	$Hg(OH)_2$	4×10^{-26}
Ag_2CrO_4	3×10^{-12}	HgS (black)	1×10^{-52}
AgI	8×10^{-17}	(red)	4×10^{-53}
AgOH	2×10^{-8}	Hg_2SO_4	7×10^{-7}
Ag_3PO_4	1×10^{-16}	$MgCO_3$	1×10^{-5}
Ag_2S	6×10^{-50}	$MgNH_4PO_4$	7×10^{-14}
AgSCN	1×10^{-12}	$Mg(OH)_2$	1×10^{-11}
Ag_2SO_4	2×10^{-5}	$Mn(OH)_2$	2×10^{-13}
$Al(OH)_3$	1×10^{-33}	MnS (pink)	3×10^{-10}
As_2S_3	†	(green)	3×10^{-13}
$BaCO_3$	5×10^{-9}		
$BaCrO_4$	1×10^{-10}	$Ni(OH)_2$	6×10^{-16}
$Ba(OH)_2$	5×10^{-3}	NiS (α)	3×10^{-19}
$BaSO_4$	1×10^{-10}	(β)	1×10^{-24}
$Be(OH)_2$	1×10^{-21}	(γ)	2×10^{-26}
Bi_2S_3	1×10^{-97}		
		$PbBr_2$	9×10^{-6}
$CaCO_3$	5×10^{-9}	$PbCO_3$	6×10^{-14}
CaC_2O_4	2×10^{-9}	$PbCl_2$	2×10^{-5}
CaF_2	3×10^{-11}	$PbCrO_4$	3×10^{-13}
$Ca(OH)_2$	4×10^{-6}	PbI_2	1×10^{-9}
$CaSO_4$	2×10^{-5}	$Pb(OH)_2$	6×10^{-16}
$Cd(IO_3)_2$	2×10^{-8}	PbS	1×10^{-28}
$Cd(OH)_2$	4×10^{-15}	$PbSO_4$	2×10^{-8}
CdS	2×10^{-28}	Sb_2S_3	2×10^{-93} ‡
$Co(OH)_2$	6×10^{-15}	$Sc(OH)_3$	2×10^{-30}
$Co(OH)_3$	3×10^{-41}	$Sn(OH)_2$	8×10^{-29}
CoS (α)	4×10^{-21}	SnS	1×10^{-25}
(β)	2×10^{-25}	$SrCO_3$	1×10^{-10}
$Cr(OH)_2$	1×10^{-17}	$SrCO_4$	2×10^{-5}
$Cr(OH)_3$	1×10^{-30}	$Sr(OH)_2$	3×10^{-4}
CuBr	5×10^{-9}	$SrSO_4$	3×10^{-7}
CuCl	2×10^{-7}		
CuI	1×10^{-12}	$TiO(OH)_2$	1×10^{-29}
$Cu(OH)_2$	1×10^{-20}		
CuS	6×10^{-36}	$V(OH)_2$	4×10^{-46}
		$V(OH)_3$	4×10^{-35}
$Fe(OH)_2$	8×10^{-16}	$VO(OH)_2$	3×10^{-24}
$Fe(OH)_3$	4×10^{-40}		
FeS	5×10^{-18}	$Zn(OH)_2$	3×10^{-17}
		ZnS (zinc blende)	2×10^{-2}
		(wurtzite)	3×10^{-22}
$Ga(OH)_3$	8×10^{-40}		

Note: For a solid M_aL_b in equilibrium with its ions in aqueous solution, $M_aL_b(s) \rightleftharpoons a\ M^{b+} + b\ L^{a-}$, the solubility product is the product of the concentrations, $K_{sp} = [M^{b+}]^a[L^{a-}]^b$, each expressed in moles per liter.

*$2AgCN(s) \rightleftharpoons Ag^+ + Ag(CN)_2^-$ $K_{sp} = 5 \times 10^{-12}$

†$1/2 As_2S_3(s) + 3H_2O \rightleftharpoons As(OH)_3 + 3/2 H_2S(g)$ $K_{sp} = 3 \times 10^{-13}$

‡$1/2 Sb_2S_3(s) + 3H_2O \rightleftharpoons Sb(OH)_3 + 3/2 H_2S(g)$ $K_{sp} = 2 \times 10^{-15}$

TABLE 7
CONCENTRATION OF SIDE-SHELF ACID AND BASE SOLUTIONS

Reagent	Formula	Molarity	Density at 25°C (g/mL)
acetic acid (glacial)	$HC_2H_3O_2$	17	1.05
acetic acid (dilute)		6	1.04
hydrochloric acid (concentrated)	HCl	12	1.18
hydrochloric acid (dilute)		6	1.10
nitric acid (concentrated)	HNO_3	16	1.42
nitric acid (dilute)		6	1.19
sulfuric acid (concentrated)	H_2SO_4	18	1.84
sulfuric acid (dilute)		3	1.18
ammonia solution (concentrated)	NH_3	15	0.90
ammonia solution (dilute)		6	0.96
sodium hydroxide (dilute)	NaOH	6	1.22

Appendices A, B, C, D

TABLE 8
STANDARD REDUCTION POTENTIALS AT 25° C

Half Reaction	Standard Potential $\epsilon°$ (Volts)	Half Reaction	Standard Potential $\epsilon°$ (Volts)
$F_2 + 2e^- \rightleftharpoons 2F^-$	2.87	$2H^+ + 2e^- \rightleftharpoons H_2(g)$	0.000
$Co^{3+} + e^- \rightleftharpoons Co^{2+}$	1.82	$O_2(g) + H_2O + 2e^- \rightleftharpoons OH^- + HO_2^-$	−0.076
$H_2O_2 + 2H^+ + 2e^- \rightleftharpoons 2H_2O$	1.77	$Pb^{2+} + 2e^- \rightleftharpoons Pb$	−0.126
$PbO_2 + 4H^+ + SO_4^{2-} + 2e^- \rightleftharpoons PbSO_4 + 2H_2O$	1.70	$CrO_4^{2-} + 4H_2O + 3e^- \rightleftharpoons 5OH^- + Cr(OH)_3(s)$	−0.13
$Ce^{4+} + e^- \rightleftharpoons Ce^{3+}$	1.61	$Sn^{2+} + 2e^- \rightleftharpoons Sn$	−0.136
$MnO_4^- + 8H^+ + 5e^- \rightleftharpoons Mn^{2+} + 4H_2O$	1.51	$Ni^{2+} + 2e^- \rightleftharpoons Ni$	−0.250
$Au^{3+} + 3e^- \rightleftharpoons Au$	1.50	$Co^{2+} + 2e^- \rightleftharpoons Co$	−0.277
$Cl_2(g) + 2e^- \rightleftharpoons 2Cl^-$	1.36	$PbSO_4(s) + 2e^- \rightleftharpoons SO_4^{2-} + Pb$	−0.356
$Cr_2O_7^{2-} + 14H^+ + 6e^- \rightleftharpoons 2Cr^{3+} + 7H_2O$	1.33	$Cd^{2+} + 2e^- \rightleftharpoons Cd$	−0.403
$MnO_2 + 4H^+ + 2e^- \rightleftharpoons Mn^{2+} + 2H_2O$	1.23	$Cr^{3+} + e^- \rightleftharpoons Cr^{2+}$	−0.41
$O_2(g) + 4H^+ + 4e^- \rightleftharpoons 2H_2O$	1.23	$2H^+(10^{-7}M) + 2e^- \rightleftharpoons H_2(g)$	−0.414
$2IO_3^- + 12H^+ + 10e^- \rightleftharpoons I_2 + 6H_2O$	1.20	$Fe^{2+} + 2e^- \rightleftharpoons Fe$	−0.44
$Br_2 + 2e^- \rightleftharpoons 2Br^-$	1.09	$S + 2e^-(1\,M\,OH^-) \rightleftharpoons S^{2-}$	−0.48
$OCl^- + H_2O + 2e^- \rightleftharpoons Cl^- + 2OH^-$	0.94	$2CO_2(g) + 2H^+ + 2e^- \rightleftharpoons H_2C_2O_4(aq)$	−0.49
$2Hg^{2+} + 2e^- \rightleftharpoons Hg_2^{2+}$	0.92	$Cr^{3+} + 3e^- \rightleftharpoons Cr$	−0.74
$Cu^{2+} + I^- + e^- \rightleftharpoons CuI$	0.85	$Zn^{2+} + 2e^- \rightleftharpoons Zn$	−0.763
$Ag^+ + e^- \rightleftharpoons Ag$	0.80	$2H_2O + 2e^- \rightleftharpoons 2OH^- + H_2(g)$	−0.828
$Hg_2^{2+} + 2e^- \rightleftharpoons 2Hg$	0.79	$SO_4^{2-} + H_2O + 2e^- \rightleftharpoons 2OH^- + SO_3^{2-}$	−0.93
$Fe^{3+} + e^- \rightleftharpoons Fe^{2+}$	0.771	$Mn^{2+} + 2e^- \rightleftharpoons Mn$	−1.18
$O_2 + 2H^+ + 2e^- \rightleftharpoons H_2O_2$	0.68	$Al^{3+} + 3e^- \rightleftharpoons Al$	−1.66
$MnO_4^- + 2H_2O + 3e^- \rightleftharpoons MnO_2 + 4OH^-$	0.59	$Mg^{2+} + 2e^- \rightleftharpoons Mg$	−2.37
$Cu^{2+} + Cl^- + e^- \rightleftharpoons CuCl$	0.566	$Na^+ + e^- \rightleftharpoons Na$	−2.714
$I_2 + 2e^- \rightleftharpoons 2I^-$	0.54	$Ca^{2+} + 2e^- \rightleftharpoons Ca$	−2.87
$Cu^{2+} + 2e^- \rightleftharpoons Cu$	0.34	$Sr^{2+} + 2e^- \rightleftharpoons Sr$	−2.89
$Hg_2Cl_2 + 2e^- \rightleftharpoons 2Hg + 2Cl^-$	0.270	$Ba^{2+} + 2e^- \rightleftharpoons Ba$	−2.90
$Hg_2Cl_2 + 2e^- \rightleftharpoons 2Hg + 2Cl^-$ (saturated KCl)	0.244	$Cs^+ + e^- \rightleftharpoons Cs$	−2.92
$AgCl + e^- \rightleftharpoons Ag + Cl^-$	0.222	$K^+ + e^- \rightleftharpoons K$	−2.925
$Cu^{2+} + e^- \rightleftharpoons Cu^+$	0.15	$Li^+ + e^- \rightleftharpoons Li$	−3.045
$Sn^{4+} + 2e^- \rightleftharpoons Sn^{2+}$	0.13		

Note: $\epsilon° = \epsilon°_{cell}$ when combined with $H_2\,(a=1) \rightarrow 2H^+\,(a=1) + 2e^-$.

TABLE 9
COMMON BUFFER SOLUTION

Buffer	Preparation
solution of pH 4 (0.050 M KHP)	Dissolve 102 g of potassium acid phthalate in 10.0 L of distilled water that has been boiled and cooled. As a preservative, add 0.6 g of thymol. The pH of this solution is 4.00-4.01 (20° -30°C).
solution of pH 5 (0.5 M HAc and 0.5 M NaAc)	Dissolve 29 mL of glacial acetic acid in 68 g of sodium acetate trihydrate and dilute to 1 L with distilled water.
solution of pH 10 (NH_3 and NH_4Cl)	Dissolve 32 g of ammonium chloride in 100 mL of water. Add 285 mL of concentrated ammonia and dilute to 500 mL.

TABLE 10
COLOR CHANGES, pH INTERVALS AND H_3O^+ CONCENTRATIONS INTERVALS OF SOME IMPORTANT INDICATORS

Indicator	pH Interval	Color Change	$[H_3O^+]$ Interval
methyl violet	0.2-3.0	yellow, blue, violet	$0.6 - 1.0 \times 10^{-3}$
thymol blue	1.2-2.8	red to yellow	$6.3 \times 10^{-2} - 1.6 \times 10^{-3}$
tropeolin 00 (orange IV)	1.3-3.0	red to yellow	$5.0 \times 10^{-1} - 1.00 \times 10^{-3}$
methyl orange	3.1-4.4	red to orange to yellow	$7.9 \times 10^{-4} - 4.0 \times 10^{-5}$
bromphenol blue	3.0-4.6	yellow to blue violet	$1.0 \times 10^{-2} - 2.5 \times 10^{-5}$
congo red	3.0-5.0	blue to red	$1.0 \times 10^{-3} - 1.0 \times 10^{-1}$
bromcresol green	3.8-5.4	yellow to blue	$1.6 \times 10^{-4} - 4.0 \times 10^{-6}$
methyl red	4.4-6.2	red to yellow	$4.0 \times 10^{-5} - 6.3 \times 10^{-7}$
litmus	4.5-8.3	red to blue	$3.2 \times 10^{-5} - 5.0 \times 10^{-8}$
bromthymol blue	6.0-7.6	yellow to blue	$1.0 \times 10^{-6} - 2.5 \times 10^{-8}$
phenol red	6.8-8.2	yellow to red	$1.6 \times 10^{-7} - 6.3 \times 10^{-9}$
cresol red	7.0-9.1	yellow to red	$1.0 \times 10^{-7} - 7.9 \times 10^{-10}$
thymol blue	8.0-9.6	yellow to blue	$1.0 \times 10^{-8} - 2.5 \times 10^{-10}$
phenolphthalein	8.0-9.8	colorless to red	$1.0 \times 10^{-8} - 1.6 \times 10^{-10}$
thymol phthalein	9.3-10.5	yellow to blue	$5.0 \times 10^{-10} - 3.16 \times 10^{-11}$
alizarin yellow R	10.0-12.0	yellow to red	$1.00 \times 10^{-10} - 1.00 \times 10^{-12}$
tropeolin 0	11.1-12.6	yellow to orange	$7.94 \times 10^{-11} - 2.51 \times 10^{-13}$
trinitrobenzene	12.0-14.0	colorless to orange	$1.00 \times 10^{-12} - 1.00 \times 10^{-14}$

Appendices A, B, C, D

Appendix C
Understanding the Spectrophotometer

DESCRIPTION OF THE INSTRUMENT

The basic components of a spectrophotometer are: a monochromator, a device which supplies light at a single wavelength for passage through the sample; a sample holder for supporting the solution in the beam of light from the monochromator; and a detector for measuring the intensity of the light after its passage through the sample.

This spectrophotometer can be used without modification to measure radiation between 340 and 600 nanometers, abbreviated nm (1 nm = 10^{-9} m). Accessories are available that extend the upper wavelength limit to 950 nm. The components are shown in Fig. C-1. *Read this material carefully* so you will have a better understanding of instrument operations.

A tungsten lamp serves as a source of radiation. White light emanating from the lamp passes through an entrance slit and is dispersed by a diffraction grating. From the dispersed beam, a narrow band containing light of similar wavelengths passes through a second slit into the sample solution. Any radiation not absorbed by the solution passes through and falls upon a phototube where its intensity is measured electronically.

The diffraction grating is a reflection grating having 1200 or more accurately spaced lines per millimeter. The white light falling on this grating is dispersed into a horizontal fan of beams ranging from ultraviolet and violet on one end to the long red and near-infrared wavelengths on the other. This spectrum of radiation falls on a wall with a narrow slit (the exit slit) cut in it. Only that portion of the spectrum falling on the slit can pass through to the sample. Turning the grating lets any desired part of the spectrum pass through the slit. The grating is turned by a knob called the wavelength selector, generally located on top of the instrument. Attached to this knob is a graduated scale which is calibrated to indicate the wavelength of light centered on the exit slit. The wavelength is given in nanometers, and the slit of the instrument passes a band of wavelengths 10 nm wide.

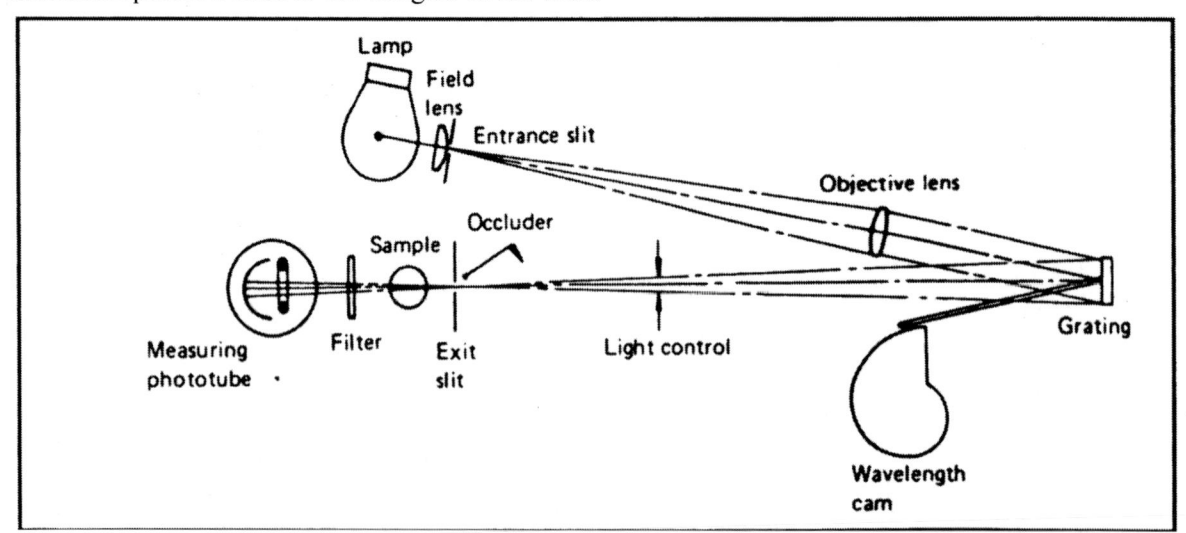

Fig. C-1. Schematic optical diagram for a single beam spectrophotometer (from Milton Roy Corporation Spectronic 21 Operating Manual).

To measure the absorbance due to a particular species in solution, you must compensate for the absorbance of the solvent and other species present in the solution. This is done after "zeroing" the instrument by adjusting the zero control so that the meter reads 0% T when there is no light reaching the phototube (no cuvette in the sample holder.) After zeroing, insert the blank solution in the instrument. The blank contains the pure solvent and all the components present in the solution except the species being measured. Rotate the light control clockwise until the meter reads 100% transmittance (100% T). *Warning: do not peg the meter.* If the sample solution is now substituted for the blank, any change in reading is due to the particular light-absorbing species in the sample, and the percent T or absorbance recorded by the instrument is a measure of the quantity of absorbing species present.

Whenever a change in wavelength is made, the 100% T must be reset with the blank, since the amount of compensation needed will vary with wavelength according to the absorption characteristics of the blank. The meter needle should go to 0% T whenever the cuvette is removed from the sample holder and the occluder drops into position. Rezeroing may be necessary because of fluctuations in the electronic components. To exclude the entrance of ambient light, close the cover on the sample holder when making a measurement.

Appendices A, B, C, D

# Appendix D
Quick View of the pH Meter-Voltmeter

DESCRIPTION OF THE pH MEASUREMENT

The pH meter is used primarily to measure the pH of aqueous solutions. The typical commercial instrument is essentially a voltmeter which measures the potential of a special type of galvanic cell in which the cell potential, ε_{cell} is sensitive to the concentration of H_3O^+ ions. A sample solution of unknown pH is incorporated into this galvanic cell and a voltage measurement is taken. Since the voltmeter scale has been calibrated to read directly in pH units (from 0 to 14), the reading corresponds to the pH of the sample solution. (Most pH meters have an auxiliary millivolt scale as well, which indicates the cell potential.)

The galvanic cell usually consists of a reference calomel electrode and an indicating glass electrode dipping into the aqueous electrolytic solution of unknown pH. The calomel electrode consists of an [Hg|Hg$_2$Cl$_2$] electrode in contact with KCl solution. The glass electrode consists of an Ag|AgCl electrode dipping into an HCl solution (usually 0.1 M) contained in a special glass bulb with a thin membrane that is permeable to H_3O^+ ions. When this electrode is dipped into a solution with a pH different from that of the HCl solution, a potential is established across the membrane. The complete cell may be represented schematically as follows.

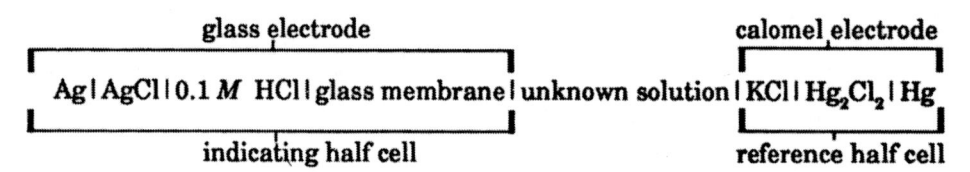

```
              glass electrode                              calomel electrode
        ┌─────────────────────────┐                   ┌──────────────────────┐
        Ag | AgCl | 0.1 M HCl | glass membrane | unknown solution | KCl | Hg₂Cl₂ | Hg
        └─────────────────────────┘                   └──────────────────────┘
              indicating half cell                          reference half cell
```

Since the glass membrane has a very high resistance, the current flowing is small and the cell voltage is a good approximation of ε_{cell}.

The observed cell potential, ε_{cell}, can be expressed as

$$\varepsilon_{cell} = \varepsilon_I - \varepsilon_R$$

where ε_I and ε_R are the potentials of the indicating and reference half cells, respectively. Applying the Nernst equation to the indicating half-cell at 25° C,

$$\varepsilon_I = \varepsilon_I^\circ - 0.059 \log [H_3O^+] = \varepsilon_I^\circ + 0.059 \, pH$$

In this equation ε_I° includes constant terms due to the activity of the [Ag|AgCl] electrode and the H_3O^+ concentration of the 0.1 M HCl solution. The pH is that of the unknown solution. Since ε_R is also constant, ε_{cell} may be expressed as

$$\varepsilon_{cell} = \text{constant} + 0.059 \, pH \text{ at } 25°\text{ C}$$

Thus, the observed cell voltage increased linearly with pH, and a difference of 1 pH unit produces a change of 0.059 V in the ε_{cell} at 25° C. Since the Nernst equation is not obeyed exactly, due in part to slight imperfections in the glass, for accurate work the meter must be *standardized* using standard buffers of precisely known pH. Also, the pH meter contains a variable resistor which can be adjusted to correct the readings when the meter is used at temperatures other than 25° C.

Appendices A, B, C, D

List of Periodic Table Elements

Name	Symbol	Number	Atomic weight	Name	Symbol	Number	Atomic weight	Name	Symbol	Number	Atomic weight
actinium*	Ac	89	(227.028)	hafnium	Hf	72	178.49	praseodymium	Pr	59	140.908
aluminum	Al	13	26.981	hassium*	Hs	108	(265)	promethium*	Pm	61	(145)
americium*	Am	95	(243)	helium	He	2	4.00260	protactinium*	Pa	91	231.036
antimony	Sb	51	121.76	holmium	Ho	67	164.930	radium*	Ra	88	(226.025)
argon	Ar	18	39.948	hydrogen	H	1	1.0079	radon*	Rn	86	(222)
arsenic	As	33	74.9216	indium	In	49	114.818	roentgenium*	Rg	111	(273)
astatine*	At	85	(210)	iodine	I	53	126.904	rhenium	Re	75	186.207
barium	Ba	56	137.33	iridium	Ir	77	192.22	rhodium	Rh	45	102.906
berkelium*	Bk	97	(247)	iron	Fe	26	55.845	rubidium	Rb	37	85.4678
beryllium	Be	4	9.01218	krypton	Kr	36	83.798	ruthenium	Ru	44	101.07
bismuth	Bi	83	208.980	lanthanum	La	57	138.905	rutherfordium*	Rf	104	(261.11)
bohrium*	Bh	107	(262)	lawrencium*	Lr	103	(260)	samarium	Sm	62	150.36
boron	B	5	10.811	lead	Pb	82	207.2	scandium	Sc	21	44.9559
bromine	Br	35	79.904	lithium	Li	3	6.941	seaborgium*	Sg	106	(263.12)
cadmium	Cd	48	112.411	livermorium	Lv	116	(293)	selenium	Se	34	78.96
cesium	Cs	55	132.905	lutetium	Lu	71	174.967	silicon	Si	14	28.0855
calcium	Ca	20	40.078	magnesium	Mg	12	24.305	silver	Ag	47	107.868
californium*	Cf	98	(251)	manganese	Mn	25	54.9380	sodium	Na	11	22.9898
carbon	C	6	12.011	meitnerium*	Mt	109	(266)	strontium	Sr	38	87.62
cerium	Ce	58	140.116	mendelevium*	Md	101	(258)	sulfur	S	16	32.065
chlorine	Cl	17	35.453	mercury	Hg	80	200.59	tantalum	Ta	73	180.948
chromium	Cr	24	51.996	molybdenum	Mo	42	95.94	technetium*	Tc	43	(98)
cobalt	Co	27	58.9332	moscovium*	Mc	115	(289)	tellurium	Te	52	127.60
copernicum	Cn	112	(277)	neodymium	Nd	60	144.242	tennessine*	Ts	117	(294)
copper	Cu	29	63.546	neon	Ne	10	20.1797	terbium	Tb	65	158.925
curium*	Cm	96	(247)	neptunium*	Np	93	(237.048)	thallium	Tl	81	204.383
darmstadtium*	Ds	110	(269)	nickel	Ni	28	58.6934	thorium*	Th	90	232.038
dubnium*	Db	105	(262.11)	nihonium*	Nh	113	(286)	thulium	Tm	69	168.934
dysprosium	Dy	66	162.500	niobium	Nb	41	92.9064	tin	Sn	50	118.71
einsteinium*	Es	99	(252)	nitrogen	N	7	14.0067	titanium	Ti	22	47.867
erbium	Er	68	167.259	nobelium*	No	102	(259)	tungsten	W	74	183.84
europium	Eu	63	151.964	oganesson*	Og	118	(294)	uranium*	U	92	238.029
fermium*	Fm	100	(257)	osmium	Os	76	190.23	vanadium	V	23	50.9415
flerovium	Fl	114	(289)	oxygen	O	8	15.9994	xenon	Xe	54	131.29
fluorine	F	9	18.9984	palladium	Pd	46	106.42	ytterbium	Yb	70	173.04
francium*	Fr	87	(223)	phosphorus	P	15	30.9738	yttrium	Y	39	88.9059
gadolinium	Gd	64	157.25	platinum	Pt	78	195.08	zinc	Zn	30	65.409
gallium	Ga	31	69.723	plutonium*	Pu	94	(244)	zirconium	Zr	40	91.22
germanium	Ge	32	72.64	polonium*	Po	84	(209)				
gold	Au	79	196.967	potassium	K	19	39.0983				

Atomic weights to 6 figures or less adapted from *Pure Appl. Chem.* 78(11), 2051-2066, 2006 and Aug 2007 release. *Elements have no stable nuclides.

General Chemistry
Quantitative and Qualitative Laboratory Experiments 3E
Russell·Bramwell·Pritchett·Reeves·Tourné·Abugri

Dr. Albert E. Russell attended UNC-Chapel Hill where obtained his Ph.D. with a concentration in organic/organometallic chemistry. While there, he was a GEM Consortium fellow supported by DuPont as well as a Sloan Scholar. He also completed an NIH postdoctoral fellowship at the University of Maryland. His current research interests are in the area of small molecule chemotherapeutics, C-O bond deconstruction, natural products for medicinal applications, natural insecticide formulation, and C-H activation of cellulosic materials. He has a serious passion for teaching, learning, and introducing innovative pedagogical methods into the classroom.

Dr. Fitzgerald B. Bramwell served as Dean of the College of Arts and Sciences at Tuskegee University; Associate Provost for Academic Research at Howard University; Vice President for Research and Graduate Studies at the University of Kentucky; and Dean of Graduate Studies and Research at Brooklyn College, CUNY. He earned his B.A. at Columbia University and M.S. and Ph. D. degrees in physical chemistry at the University of Michigan. His research interests include organotin chemistry, multidimensional organic conductors, and photoexcited radicals.

Dr. Gregory Pritchett is a former Associate Dean of the College of Arts and Sciences and former Chair, Department of Chemistry at Tuskegee University. His research interests include the investigation of structural changes in isolated type I blue copper proteins and natural sources of dyes and indicators for commercial applications. He earned his B.S. and M.S. degrees in chemistry at Tuskegee University and his Ph. D. degree in biophysical chemistry at Georgia State University.

Dr. Melissa S. Reeves received her Ph.D. from Indiana University, Bloomington. Her current research focuses on simulating interactions between polymers and inserted nanoparticles, particularly the effect of the particle on the macroscopic properties of the composite. Reeves' teaching interests and innovations have primarily related to the teaching of Physical Chemistry. She is active in writing materials for the POGIL-PCL project.

Professor Marilyn Tourné received her Ph.D. from the University of Florida, Gainesville. Her current research focuses on explosive analysis using mass spectrometry (MS). Dr. Tourné is also interested in the development of novel analytical platforms to increase sensitivity, selectivity, and specificity of detection and deterministic identification of target analytes. Her academic interests are related primarily to teaching General and Analytical Chemistry, and include exploring different teaching methodologies such as guided-inquiry based models, blended, and fully online formats.

Professor Daniel A. Abugri received his B.Sc. (Hons.) in Applied Chemistry with Enivironmental Science from the University for Development Studies, Navrongo, Ghana, and M.Sc. in Chemistry with concentration in Analytical Biochemistry, and Ph.D. in Integrative Biosciences with concentration in Biochemistry and Molecular Parasitology from Tuskegee University. Dr. Abugri's current research encompasses green chemistry, chemical and biological education, drug discovery from medicinal plants and natural products for treatment of cancer, malaria, toxoplasmosis, fungal and bacterial diseases. He also has a passion for introducing undergraduate students to research and teaching them sound scientific principles. Dr. Abugri currently holds a joint Assistant Professorship between the Chemistry and Biology departments at Tuskegee University.

ISBN: 978-1-942465-02-7